BOUNDARY ELEMENT PROGRAMMING IN MECHANICS

Nonlinear stress analysis is an essential feature in the design of such diverse structures as aircraft, bridges, machines, and dams. Computational techniques have become vital tools in dealing with the complex, time-consuming problems associated with nonlinear stress analysis. Although finite element techniques are widely used, boundary element methods (BEM) offer a powerful alternative, especially in tackling problems of three-dimensional plasticity.

This 2002 book describes the application of BEM in solid mechanics, beginning with basic theory and then expalining the numerical implementation of BEM in three-dimensional nonlinear stress analysis. In, addition, the authors have developed state-of-the-art BEM source code, available on www.cambridge.org/9780521773591. The main topics covered are: (a) development of the partial differential equations that describe the elasto-plastic (flow theory) mechanics of continua, (b) formulation of the boundary integral equations for elasto-plasticity, (c) description of the numerical algorithms in the implementation of the boundary element method, (d) methods for evaluating the singularities and solving the nonlinear system equations, (e) description of the source code, and (f) presentation of benchmarks and applications.

This book will be especially useful to stress analysts in industry, research workers in the field of computational plasticity, and postgraduate students taking courses in engineering mechanics.

Xiao-Wei Gao is Research Associate, Department of Mechanical and Aerospace Engineering, Arizona State University, Tempe. He was formerly Associate Professor, Institute of Applied Mechanics, Ningxia University, People's Republic of China. Dr. Gao is an authority on nonlinear boundary element analysis and is currently developing BEM techniques for commercial aerospace applications.

Trevor G. Davies is Senior Lecturer, Glasgow University. Dr. Davies's primary expertise is in numerical analysis of nonlinear and dynamic soil-structure interaction using boundary element methods. He was one of the first researchers to tackle the problem of 3D nonlinear boundary' element analysis during the late 19708. He has published extensively and co-edited the book *Boundary Element Techniques in Geomechanics*.

BOUNDARY ELEMENT
PROGRAMMING
IN MECHANICS

XIAO-WEI GAO

Arizona State University

TREVOR G. DAVIES

Glasgow University

CAMBRIDGE
UNIVERSITY PRESS

CAMBRIDGE UNIVERSITY PRESS
Cambridge, New York, Melbourne, Madrid, Cape Town, Singapore,
São Paulo, Delhi, Dubai, Tokyo, Mexico City

Cambridge University Press
The Edinburgh Building, Cambridge CB2 8RU, UK

Published in the United States of America by Cambridge University Press, New York

www.cambridge.org
Information on this title: www.cambridge.org/9781107400252

First published 2002
First paperback edition 2011

A catalogue record for this publication is available from the British Library

Library of Congress Cataloguing in Publication Data

Gao, Xiao-Wei, 1960-
    Boundary element programming in mechanics / Xiao-Wei Gao, Trevor G. Davies.
        p. cm.
    Includes bibliographical references.
    ISBN 0-521-77359-8
    1. Boundary element methods - Data processing. 2. Mechanics, Applied.
    3. Computer programming. I. Davies, Trevor G., 1954- II. Title.
    TA347.B69 G36 2001
    620.1'001'51535 - dc21   2001025613

ISBN 978-0-521-77359-1 Hardback
ISBN 978-1-107-40025-2 Paperback

To Fang Wang and Eileen

Contents

Preface

The essence of this book is the computer code. In these pages, we bring three-dimensional nonlinear code for boundary element analysis of solid continua into the public arena for the first time. By this means, we hope to stimulate greater use of this elegant analysis in practice and to spur its further development in academia.

The book is aimed at three groups of people: (1) practitioners (stress analysts) in diverse engineering fields, such as mechanical, aeronautical, and civil engineering, (2) graduate students in these disciplines, and (3) numerical analysts (researchers) in academia and elsewhere.

To the first group, the software described in this book offers a powerful alternative to existing methods of analysis (e.g., finite elements) and can be used, for example, for validating numerical results obtained by such methods. Moreover, for some three-dimensional applications, the software will likely become the method of choice. For the second group, the book demonstrates how theory is translated into practical software. This, we hope, will yield a clearer, more concrete, exposition of the subject than a lengthy abstract description could possibly achieve. Finally, we hope that researchers will be encouraged to take our work forward and to uncover new facets of this remarkable technique.

Meeting the expectations of this readership has not been easy and some compromises have been necessary. Others must judge whether out attempt to steer this new path has avoided the pitfalls of traversing too much of the well-worn highways of solid mechanics or whether we have ventured too far into uncharted territory. We have also been at pains to take the shortest practical route to our goal, when perhaps we could have lingered more often to survey the scene and to describe it in more detail for those who would follow us. However, we have tried to avoid the temptation to take too many shortcuts in the hope that others, like us, will take pleasure from the journey itself.

Historical Note

Central to this book is the fundamental singular solution of solid mechanics, known as Kelvin's solution (Thomson, 1848*), a theoretical discovery that is uniquely associated with Glasgow University. A small blue plaque fixed to one of the town houses (No. 11) in "Professor's Square" on the main campus reads: "William Thomson (Baron Kelvin of Largs), Physicist, lived here[†] 1870–1899." We dare to suppose that this giant of nineteenth-century physics, a practical man nonetheless, would approve of our efforts to apply his elegant results to engineering problems in the twenty-first century. Although Irish born, he took the name Largs from the coastal resort, some forty miles distant from Glasgow, where he built a mansion. Close to the University flows the Kelvin, a major tributary of the Clyde. His brother, James, was the third incumbent of the Regius Chair of Civil Engineering and Mechanics at Glasgow.

Acknowledgments

The authors have been privileged to count among their colleagues many who have made significant contributions to the canon of boundary element research. They have shaped our thinking in diverse ways, and our efforts would be diminished without their contribution to our work. We are glad to have this opportunity to thank them (albeit anonymously) here. Two individuals who have particularly influenced our careers deserve special mention: Prof. Y. R. Zheng first directed the senior author (XWG) to this subject and encouraged him to write a book on nonlinear boundary element methods. Similarly, the second author (TGD) cordially acknowledges his indebtedness to Prof. P. K. Banerjee, internationally renowned for his pioneering BEM research, who first introduced him to boundary element methods and illuminated many of its subtleties. His direct influence is apparent in our citations of his innovative work on nonlinear analysis. We are also much indebted to Prof. G. Beer, Dr. P. Bhatt, and Mr. C. Duenser for many helpful comments on the manuscript. Lastly, we are extremely grateful to Ms. F. Padgett and her colleagues at Cambridge University Press for their patience and encouragement.

November 2000   *Xiao-Wei Gao and Trevor G. Davies*

---

* This short paper, submitted in December 1847, was published just two months later. It generalizes the results of his paper of the preceding year in which he treated the special "incompressible" case.
† William Thomson was appointed to the Chair of Natural Philosophy at Glasgow University in 1846, a post he retained until his retirement in 1899. He was knighted in 1862 for services to ocean telegraphy and was raised to the peerage (thus earning the right to be styled Lord Kelvin) thirty years later. Following his death in 1907, he was laid to rest in Westminster Abbey; his tomb lies alongside that of Sir Isaac Newton.

Legal Matters

Read this section if you plan to use the programs in this book on a computer.

Disclaimer of Warranty

We make no warranties, express or implied, that the programs contained in this volume are free of error or are consistent with any particular standard of merchantability or that they will meet your requirements for any particular application. They should not be relied on for solving a problem whose incorrect solution could result in injury to a person or loss of property. If you do use the programs in such a manner, it is at your own risk. The authors (X.-W. Gao & T. G. Davies) and the publisher disclaim all liability for direct or consequential damages resulting from your use of the programs.

License Information

The programs described in this volume are not in the public domain – the authors (X.-W. Gao & T. G. Davies) retain ownership and copyright and exclusively reserve all rights to the software. In countries where assertion of the right to be identified as the author is required for copyright purposes, the authors assert their right to be recognized as the authors and owners of these programs. The authors also assert their moral rights to ownership of these programs. If you are the individual owner of a copy of this book, you are permitted to use the programs for personal and noncommercial use only. You are not permitted to transfer or distribute the programs in any format to any other person. Publication of results, in journal articles or conference proceedings, obtained using these programs (or using software based substantially on these programs) must include a citation of this volume. Instructors at accredited educational establishments who have adopted this book for a course may make available to their students electronic copies of the object code only (not source code) of the programs for use in that course, but only for the duration of the course. Commercial use of these programs, or software based substantially on these programs, is expressly prohibited unless a license for this purpose has been granted by the authors (X.-W. Gao & T. G. Davies). Further information, including program updates, may be obtained from the authors at the following address: http://www.civil.gla.ac.uk/bemech.

Trademarks

Pentium is a trademark of Intel Corporation.
Windows is a trademark of Microsoft Corporation.

LINEAR PROBLEMS

CHAPTER ONE

Introduction

1.1 Introduction

Differential equations of many types have been solved by using boundary element methods (BEM) – a numerical technique that has come of age. In this book, we describe how boundary element methods may be applied to problems in solid mechanics, governed by the well-known equations of elasticity and elasto–plasticity. In this field, BEM offers many advantages, and admittedly some disadvantages, over its rivals (e.g., the finite element and finite difference methods). It scores particularly well, for example, in three-dimensional applications, and in the fields of fracture mechanics and continuum dynamics, for reasons that will become apparent later.

If you are embarking on a study of boundary element methods for the first time, you may find the theoretical and numerical formulations somewhat daunting at first. But what soon emerges is the elegance of the method and its ready translation into program code. Nevertheless, for clarity, but perhaps at the cost of some repetition, we have felt it necessary to divide the book into two distinct parts: The first part deals with linear analysis in solid mechanics, whereas the second part deals with nonlinear analysis. This division is intended to benefit both novices and experts.

The structure of both parts of the book is similar: first, we outline the relevant solid mechanics theory and then show how the resulting governing differential equations are transformed into integral equations. These latter equations, which provide the basis of the boundary element method, are then expressed in discrete (numerical) form. The program code is then described, and examples of applications (to simple benchmarks, as well as to more challenging practical problems) are presented.

1.2 A Note on Programming

The program code BEMECH is written in Fortran90, although it also contains constructs from earlier dialects. A description of this language would be out of place here, but a readable account of it, aimed at engineers and scientists, has been published by Smith (1995). We make no apology for clinging to this "obsolete" programming language (as some information technology professionals have described it), in these days of "object-oriented" programming. On the contrary, while reports of Fortran's imminent demise (for decades past) have been greatly exaggerated, most engineers know that for computationally intensive "formula translation" its efficiency is second to none, however measured. We concede that, historically, there are a few areas where Fortran has some shortcomings: For example in (a) input–output formatting, (b) dynamic array allocation, (c) intersubroutine communication, and (d) statement labeling. However, the language has evolved to counter its deficiencies in these areas.

(a) Input–Output Formatting. While pre- and postprocessing of data are vital elements of analysis (Opriessnig & Beer, 1999), it is probably not sensible to embed them in the computational engine. Graphical manipulation of data, possibly via a Windows interface, is certainly better constructed using other languages (and software tools) than Fortran, but these are outside the scope of this book. Our experience is that it is best to assemble these pre- and postprocessing elements and the computational engine within an overarching program structure, which launches each in turn. In this way, one can separate the nonportable elements from the "universal" Fortran engine when one switches platforms.

(b) Dynamic Array Allocation. Earlier versions of Fortran did not support dynamic array allocation, although it was always possible to minimize this inconvenience by exploiting various devices, such as subroutine argument lists. Now, Fortran90 (F90) overcomes this problem by means of the ALLOCATE statement. Moreover, F90 simplifies the task of reusing (conserving) memory, by virtue of the DEALLOCATE statement. For programming clarity, we make only sparing use of the latter.

(c) Intersubroutine Communication. In earlier versions of Fortran, it was necessary to pass variables between subroutines via argument lists and COMMON statements. These laborious and error-prone methods have now been swept away with the advent of the MODULE statement. A common "module" now forms the hub of an efficient communications network between subroutines. Thus, at a stroke, program code is substantially reduced in length and complexity.

(d) Statement Labeling. Statement labeling (numbering) in Fortran allows programmers great freedom but some have abused this facility and constructed spaghetti-like codes that loop into impenetrable knots. In F90, statement labeling can be dispensed with altogether by making use of the DO–ENDDO and IF–ELSE–ENDIF structures. However, we disagree with those who eschew statement labeling altogether: Among other things, statement labels perform the valuable function of identifying the end-points of nested DO-LOOPS. Consequently, our policy is to retain statement labels where they clarify program structure (e.g., nested DO-LOOPS) and omit them elsewhere. Of course, we number statements in serial order. This function (of statement labels) is recognized in Fortran90, by the introduction of DO-LOOP (descriptive) labels, but we still prefer statement labels.

1.3 Mathematical Preliminaries

Cartesian tensor (indicial) notation is used very widely in the boundary element literature, and for very good reason. Although it may be perplexing at first sight, it is immeasurably superior to the alternatives (e.g., algebraic notation), in terms of brevity, ease of manipulation, and translation into program code. For those unfamiliar with it, we illustrate here the basic rules, using a few examples. Working through these examples, the novice will soon find that the notation becomes transparent. Consider, for example, the tensor form of Hooke's law:

$$\sigma_{ij} = \lambda \delta_{ij} \varepsilon_{kk} + 2G\varepsilon_{ij} \tag{1.1}$$

This equation, as we shall see later, describes the relationship between stresses σ and strains ε for a linearly elastic isotropic material, characterized by the elastic constants λ and G. Since there are, in general, nine components of stress and nine components of strain, this single tensor equation is equivalent to nine algebraic equations. The notation for the Cartesian components of stress (σ_{xx}, σ_{xy}, σ_{xz}, σ_{yx}, etc.) is condensed by first relabeling the x, y, and z axes by the numerals 1, 2, and 3 and then representing the subscripts by the Latin letters i and j, which we allow to range from 1 to 3, in turn. Thus, we interpret the single tensor equation as the set of nine equations

$$\begin{aligned}
(i = 1, j = 1) \qquad &\sigma_{11} = \lambda \delta_{11} \varepsilon_{kk} + 2G\varepsilon_{11} \\
(i = 1, j = 2) \qquad &\sigma_{12} = \lambda \delta_{12} \varepsilon_{kk} + 2G\varepsilon_{12}
\end{aligned} \tag{1.2}$$

etc.

One of the most important characteristics of tensor equations is that the same subscripts must appear in each term of the equation; that is, the equation must be homogenous (a feature similar to dimensional similarity, in some respects). This characteristic may be somewhat obscured by the presence of "repeated subscripts," such as occurs here in the term ε_{kk}. By convention, repeated subscripts imply summation of all terms in the range (thus, ε_{kk} signifies $\varepsilon_{11} + \varepsilon_{22} + \varepsilon_{33}$). Thus, we could have written the tensor equation as

$$\sigma_{ij} = \lambda \delta_{ij} (\varepsilon_{11} + \varepsilon_{22} + \varepsilon_{33}) + 2G\varepsilon_{ij} \tag{1.3}$$

Evidently, this equation is homogenous: use of the repeated subscript convention merely allowed us to write it more compactly. The homogenous nature of tensor equations can also be obscured by repeated subscripts of the following type (now spanning over two or more elements of a term):

$$t_i = \sigma_{ij} n_j \tag{1.4}$$

which is interpreted as the three equations

$$\begin{aligned}
(i = 1) \qquad &t_1 = \sigma_{11} n_1 + \sigma_{12} n_2 + \sigma_{13} n_3 \\
(i = 2) \qquad &t_2 = \sigma_{21} n_1 + \sigma_{22} n_2 + \sigma_{23} n_3 \\
(i = 3) \qquad &t_3 = \sigma_{31} n_1 + \sigma_{32} n_2 + \sigma_{33} n_3
\end{aligned} \tag{1.5}$$

Here again, the repeated subscript within a single term implied an addition. Scalar products can be represented, for example, by equations of the form $w = f_i s_i$. Such repeated subscripts are sometimes referred to as "dummy" subscripts, since they have no global significance. That is to say, the choice of subscript is arbitrary.

Turning back once more to Hooke's law, we employed a tensor (δ_{ij}) that, as yet, we have not defined. This special symbol, called the Kronecker delta, has wide applications and is defined as follows:

$$\begin{aligned}
\delta_{ij} &= 1 \quad \text{when } i = j \\
\delta_{ij} &= 0 \quad \text{when } i \neq j
\end{aligned} \tag{1.6}$$

In matrix notation, its equivalent is the identity matrix $[I]$. Further, the Kronecker delta allows us to condense equations via substitutions such as $w_i = \delta_{ij} w_j$. Also useful is the alternating (or permutation) tensor e_{ijk}, which is defined as follows:

$$\begin{aligned}
e_{ijk} &= 0 \quad \text{when any two subscript values are equal} \\
e_{ijk} &= 1 \quad \text{when the subscript values are in cyclic order} \\
e_{ijk} &= -1 \quad \text{when the subscript values are in anticyclic order}
\end{aligned} \tag{1.7}$$

Usually, the alternating tensor arises when one wishes to calculate the determinant of a matrix. For example, for a matrix $[D]$ of rank three, $\det [D] = e_{ijk} D_{1i} D_{2j} D_{3k}$.

Tensor notation is also employed to represent the spatial differentiation of variables. For example, the partial differential of the y-component of displacement (u_y), with respect to the spatial direction x, can be represented, using the "comma" notation, as

$$\frac{\partial u_y}{\partial x} = u_{y,x} \tag{1.8}$$

In tensor notation, the entire family of such terms (containing nine members) can be represented by the single expression $u_{i,j}$. Second-order differentiation can be represented in a similar fashion. For example, consider the single term

$$\frac{\partial^2 u_z}{\partial x \partial z} = u_{z,xz} \tag{1.9}$$

The entire family of such terms (containing twenty-seven members) can be represented by the single term $u_{i,jk}$. Thus, for example, Navier's governing differential equations of elasticity can be rendered as

$$G u_{i,jj} + (\lambda + G) u_{j,ji} + b_i = 0 \tag{1.10}$$

This single tensor equation is equivalent to three algebraic equations (for $i = 1$–3 in turn), each containing seven terms. These arise because the repeated subscript j in the first tensor term is to be interpreted as an addition of three algebraic terms (e.g., for $i = 2$, $u_{i,jj} = u_{y,xx} + u_{y,yy} + u_{y,zz}$), and similarly in the second tensor term (e.g., for $i = 2$, $u_{j,ji} = u_{x,xy} + u_{y,yy} + u_{z,zy}$). Although it is clear that tensor notation is compact, perhaps more important is the fact that tensor equations can be manipulated far more easily than their algebraic equivalents, as later examples will demonstrate.

We now gather together some results that will be drawn on throughout the book. In the boundary element method, we frequently need to calculate the distance r between a so-called source point p and a field point q. The Cartesian components r_i of this distance are

$$r_i = x_i(q) - x_i(p) \tag{1.11}$$

where $x_i(q)$ and $x_i(p)$ denote the ith component of the coordinates of the field point and the source point, respectively. So, for example, $r_y = y(q) - y(p)$. Using Pythagoras's theorem, and the summation convention, we can write

$$r = (r_i r_i)^{1/2} \tag{1.12}$$

With respect to the field point q, the spatial derivatives of r are

$$r_{,i} = \frac{\partial r}{\partial x_i(q)} \tag{1.13}$$

Using the chain rule of differentiation, we obtain

$$
\begin{aligned}
r_{,i} &= \frac{\partial r}{\partial r^2} \frac{\partial r^2}{\partial r_i} \frac{\partial r_i}{\partial x_i(q)} \\
&= \frac{1}{2r}(2r_i)1 \\
&= r_i/r
\end{aligned}
\tag{1.14}
$$

Of course, this result can also be derived using algebraic notation and the reader may care to demonstrate that, for example, $\partial r/\partial y(q) = r_y/r$. Both the terms r_i and $r_{,i}$ are used extensively in boundary element formulations and it is important to distinguish carefully between them. There will also be instances where spatial derivatives of r with respect to the load point p will be needed as, for example, in the calculation of stresses in Chapter Four. From the definition of r, it should be clear that $r_{,i(q)} = -r_{,i(p)}$.

As well as tensors, we will need to make use of certain other mathematical devices (Dirac delta function, Gauss's theorem, integration by parts, vector algebra, etc.). For the sake of completeness and convenience, we briefly describe these here. Those who are unfamiliar with these techniques should be reassured by the fact that they are, generally speaking, extensions of fairly simple ideas in engineering mathematics.

The Dirac delta function $\delta(x, p)$, sometimes referred to as an impulse function, is used to represent a quantity that has a point singularity (at $x = p$) but is nevertheless associated with a finite value. An obvious and relevant example is the notion of a point force (Fig. 1.1).

In practice, a point force is a convenient fiction but in engineering analysis it is a fiction that is useful to preserve. Because the Dirac delta function must be zero

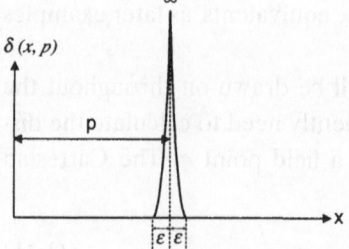

Figure 1.1: The Dirac delta function.

everywhere (in x), except at the singularity, and because its integral must assume a finite value (conveniently unity), we obtain

$$\int_{-\infty}^{\infty} \delta(x,p)dx = \int_{p-\varepsilon}^{p+\varepsilon} \delta(x,p)dx$$

$$= 1 \tag{1.15}$$

where p is the singular point and ε denotes a vanishingly small radius of integration around this singularity. Another use of the Dirac delta function exploits its capacity to isolate the value of a continuous function $f(x)$ at a specific point; thus,

$$\int_{-\infty}^{\infty} f(x)\delta(x,p)dx = \int_{p-\varepsilon}^{p+\varepsilon} f(x)\delta(x,p)dx$$

$$= f(p) \tag{1.16}$$

The same result is obtained if the integration limits are finite (a and b, say), provided that that the singularity lies in the domain of integration (i.e., $a < p < b$). Naturally, the Dirac delta function need not be defined in terms of a single variable, and more generally the variable x is understood to be the coordinates x_i and the integration is carried out over three (or two) dimensions.

Turning now to Gauss's theorem, we simply quote the well-known result (e.g., Fung, 1965) that for any continuously differentiable function F defined over some domain, the integral of its spatial derivatives over that domain can be expressed in terms of a boundary integral. More specifically, Gauss's theorem states that

$$\int_{\Omega} \frac{\partial F}{\partial x_i} d\Omega = \int_{\Gamma} F n_i d\Gamma \tag{1.17}$$

where Ω is the problem domain, Γ is the boundary (Fig. 1.2), and n_i is the ith component of the unit outward normal vector.

The reduction of certain domain integrals to surface integrals is, of course, central to the boundary integral method. The method of "integration by parts" is another basic mathematical technique that is used, often in conjunction with Gauss's theorem, for this purpose. Integration by parts is itself little more than the inverse process to differentiation of a product of two functions.

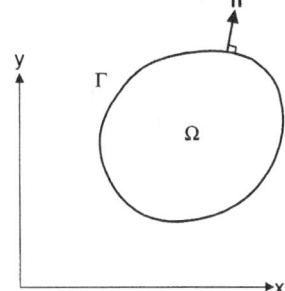

Figure 1.2: Definition sketch of problem domain.

Differentiation of the product of two functions (f and g), with respect to x_i, yields

$$\frac{\partial(fg)}{\partial x_i} = f\frac{\partial g}{\partial x_i} + g\frac{\partial f}{\partial x_i} \qquad (1.18)$$

Rearranging the terms, and integrating over a domain Ω, yields the integration by parts statement:

$$\int_\Omega g\frac{\partial f}{\partial x_i}d\Omega = \int_\Omega \frac{\partial(gf)}{\partial x_i}d\Omega - \int_\Omega f\frac{\partial g}{\partial x_i}d\Omega \qquad (1.19)$$

Now using Gauss's theorem (Eq. 1.17) we obtain

$$\int_\Omega g\frac{\partial f}{\partial x_i}d\Omega = \int_\Gamma gfn_i d\Gamma - \int_\Omega f\frac{\partial g}{\partial x_i}d\Omega \qquad (1.20)$$

By this means, it is often possible to reduce certain domain integrals appearing in the formulation of the boundary element method more amenable forms. For example, consider the one-dimensional second-order to differential equation

$$\frac{\partial^2 f}{\partial x^2} = 0 \qquad (1.21)$$

Now, let $g(x,p)$ be the fundamental solution of this equation, that is, the distribution of f that arises in response to a unit point source disturbance at p. If $g(x,p)$ is the fundamental solution, it is implicitly assumed that no boundary conditions are applied at any finite distance from p. That is, the domain is infinite. It should be noted that Eq. (1.21) is analogous to the governing differential equation for potential problems (such as heat flow) and is not too distant from those describing the mechanics of solids. Formally, $g(x,p)$ is the solution of the equation

$$\frac{\partial^2 g}{\partial x^2} + \delta(x,p) = 0 \qquad (1.22)$$

where $\delta(x,p)$ is the Dirac delta function defined earlier. While the point source term here is merely a mathematical device, in solid mechanics problems it can be identified with a point force. Using analytical methods (which we need not delve into here) the fundamental solution is found to be

$$g(x, p) = \frac{1}{2}(c - |r|) \tag{1.23}$$

where c is an arbitrary constant and r is the distance between the field point and source (load) point; that is, $r = x - p$. We now multiply Eq. (1.21) by g and integrate by parts. From Eq. (1.20), we obtain

$$\int_\Omega g \frac{\partial^2 f}{\partial x^2} d\Omega = \int_\Gamma g \frac{\partial f}{\partial x} n_x d\Gamma - \int_\Omega \frac{\partial f}{\partial x} \frac{\partial g}{\partial x} d\Omega$$
$$= 0 \tag{1.24}$$

Applying the method of integration by parts again to the last term, we obtain

$$\int_\Omega g \frac{\partial^2 f}{\partial x^2} d\Omega = \int_\Gamma g \frac{\partial f}{\partial x} n_x d\Gamma - \int_\Gamma f \frac{\partial g}{\partial x} n_x d\Gamma + \int_\Omega f \frac{\partial^2 g}{\partial x^2} d\Omega$$
$$= 0 \tag{1.25}$$

Since the domain of integration is one dimensional, it is convenient to treat it as a cylinder of cross-sectional area A oriented along the x-axis. Thus, $d\Omega = A dx$ and $d\Gamma = dA$. To make matters concrete, let the lower and upper limits of integration be (a, b), respectively. At the lower limit, $n_x = -1$, while at the upper limit, $n_x = +1$. Further, provided that $a < p < b$, the functions f and g and their first derivatives are uniquely defined at the limits. Equation (1.25) thus simplifies to

$$\int_a^b g \frac{\partial^2 f}{\partial x^2} dx = \left| g \frac{\partial f}{\partial x} - f \frac{\partial g}{\partial x} \right|_a^b + \int_a^b f \frac{\partial^2 g}{\partial x^2} dx$$
$$= 0 \tag{1.26}$$

Now we invoke Eq. (1.22) and obtain

$$\left| g \frac{\partial f}{\partial x} - f \frac{\partial g}{\partial x} \right|_a^b - \int_a^b f(x)\delta(x, p)dx = 0 \tag{1.27}$$

Finally, using the integral property of the Dirac delta function, this equation simplifies (after rearrangement) to

$$f(p) = \left| g \frac{\partial f}{\partial x} - f \frac{\partial g}{\partial x} \right|_a^b \tag{1.28}$$

This result is extremely significant. It is an expression for the function f at any interior point p in terms of boundary values only. The function g and its derivative $(\partial g/\partial x)$ are known functions, while f and its derivative $(\partial f/\partial x)$ are boundary conditions. Thus, by integrating by parts, and using the property of the Dirac delta function, we have been able to transform a differential equation into a boundary

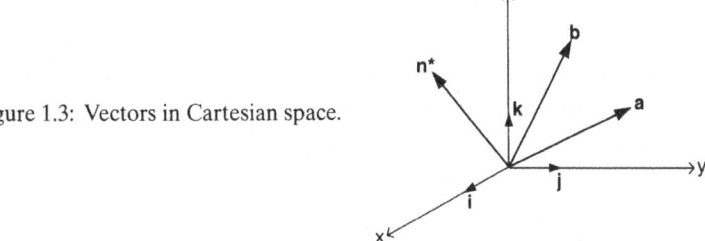

Figure 1.3: Vectors in Cartesian space.

integral equation. This is the essence of the boundary element method, even if normally one has to deal with rather more complicated differential equations, and in two (or three) dimensions. Once this technique (Eqs. 1.21–1.28) is understood, the development of the boundary integral equations in Chapters Three and Eight should not cause much difficulty.

We now turn to some mathematics related to the geometry of surfaces. First, we admit that some results relating to transformations between coordinate systems are quoted in this book without proof. (Proofs may be found in standard texts on differential geometry, e.g., Burke, 1985). However, to apply these results, some basic operations in vector algebra are necessary and it is these that we now describe. In vector algebra, we define a right-handed system of orthogonal axes (x, y, z) and a system of unit length vectors ($\mathbf{i}, \mathbf{j}, \mathbf{k}$) directed along these axes, as shown in Fig. 1.3. With reference to this figure, let $\mathbf{a}$ and $\mathbf{b}$ be two vectors, with end-points at the coordinates (a_1, a_2, a_3) and (b_1, b_2, b_3), respectively. Thus

$$
\begin{aligned}
\mathbf{a} &= a_1\mathbf{i} + a_2\mathbf{j} + a_3\mathbf{k} \\
\mathbf{b} &= b_1\mathbf{i} + b_2\mathbf{j} + b_3\mathbf{k}
\end{aligned}
\tag{1.29}
$$

The scalars a_1, a_2, etc. are the (orthogonal) components of the vector. The magnitude (length) of these vectors is determined from Pythagoras's theorem. Thus, for example,

$$
|\mathbf{a}| = \sqrt{a_1^2 + a_2^2 + a_3^2}
\tag{1.30}
$$

The normal vector $\mathbf{n}^*$ to the plane containing the vectors $\mathbf{a}$ and $\mathbf{b}$ can be determined by calculating the so-called vector cross-product:

$$
\mathbf{n}^* = \mathbf{a} \times \mathbf{b}
\tag{1.31}
$$

which may be evaluated by calculating the co-factors of the vector cross-product matrix. Thus,

$$
\mathbf{n}^* = \begin{vmatrix} \mathbf{i} & \mathbf{j} & \mathbf{k} \\ a_1 & a_2 & a_3 \\ b_1 & b_2 & b_3 \end{vmatrix}
\tag{1.32}
$$

or, in explicit form, expanding by the first row, we obtain

$$
\mathbf{n}^* = \mathbf{i}(a_2b_3 - a_3b_2) + \mathbf{j}(-a_1b_3 + a_3b_1) + \mathbf{k}(a_1b_2 - a_2b_1)
\tag{1.33}
$$

The magnitude of this vector $|\mathbf{n}^*|$ can be determined using Pythagoras's theorem, as exemplified by Eq. (1.30). Its "sense" (since $\mathbf{n}^*$ may be directed in either one of two opposite directions) is determined from the "right-hand rule." For example, the vector product $\mathbf{i} \times \mathbf{j}$ is directed along $\mathbf{k}$, whereas the vector product $\mathbf{j} \times \mathbf{i}$ is directed in the opposite direction (along $-\mathbf{k}$). Often, the unit normal vector $\mathbf{n}$ will be required, and this can be obtained by simply scaling the normal vector obtained from the vector cross-product:

$$\mathbf{n} = \mathbf{n}^*/|\mathbf{n}^*| \tag{1.34}$$

This equation implies that each component of the unit normal vector is scaled from the corresponding component in $\mathbf{n}^*$. The result of this operation is the equation

$$\mathbf{n} = n_1\mathbf{i} + n_2\mathbf{j} + n_3\mathbf{k} \tag{1.35}$$

where the components (n_1, n_2, n_3) are often referred to as the "direction cosines" of the vector.

1.4 Historical Sketch

It is a commonplace that the burgeoning scientific literature is both a blessing and a curse. Thus, in the very brief review that follows, we cite only a few representative papers from those that have crossed our desks. Of course, we could have cited many other equally worthy publications. Thus, the general reader should not be under any misapprehension that this field has attracted few workers. Equally, in this light, we ask for the indulgence of those researchers whose contributions have not been recorded here.

1.4.1 Approximate Methods

The role of boundary element methods for the solution of boundary value problems in solid mechanics should be viewed in the context of the principal alternative approximate methods, namely, the finite difference and finite element methods. The finite difference method (Southwell, 1946) is based on the direct discretization of the governing differential equations at sufficiently many points (nodes) in the domain of the problem and results in a narrow-banded set of system equations. The drawbacks of this method are the large number of nodes required to obtain accurate solutions and the difficulty in dealing with boundary conditions. The finite element method (Zienkiewicz, 1977; Owen & Hinton, 1980) has been outstandingly successful in applications to a very wide range of problems. In this method, the domain of the body is divided into elements, and distributions of the physical variables (e.g., displacements, potentials) in an element are entirely determined in terms of their local (nodal) values, via interpolation (shape) functions. The resulting system equations, involving the nodal values as unknowns, are banded and often symmetric. Because material properties are specified at element level, the finite element method can deal with inhomogeneous materials as easily as single-material problems. This feature makes it very versatile and consequently it has become the dominant numerical method. There are, however, many classes of problems (principally, infinite or semi-infinite problems and fracture analysis and dynamics) for which the finite element method is not ideally suited and more efficient techniques, based on boundary element methods, are attractive options. The boundary element method (Brebbia, 1978; Banerjee, 1994) has emerged relatively recently as a powerful numerical method of analysis of continuum problems, although its roots lie much earlier in the mathematical theory of integral equations, largely associated with the work of Fredholm (1905) and, later, Mikhlin (1965). In this technique, the governing differential equations are transformed into boundary integral equations, either by means of a reciprocal identity or, more generally, by weighted residual and integration by parts techniques. The so-called fundamental solutions (Green's functions) of the governing differential equations play a pivotal role in these integral equations. The primary unknowns involve variables defined over the boundary (only) of the body, which, in effect, reduces the dimensionality of the problem by one. The advantages, in terms of data generation and processing, particularly in three dimensions, can be substantial. Further, the use of fundamental solutions implicitly incorporates the boundary conditions at infinity, obviating the necessity to curtail such domains artificially. In fracture mechanics (e.g., Aliabadi, 1997) and dynamics (e.g., Dominguez, 1993), the constraints imposed by numerical interpolations within the domain (as employed in finite element methods) are absent, yielding much better accuracy. These, and other, advantages have spurred the use of boundary element methods in a wide range of applications.

1.4.2 BEM in Solid Mechanics

The boundary element method has evolved along two closely linked, but distinct, branches, based on whether the formulation can be classified as "indirect" or "direct." In the indirect formulation, the integral equations are usually expressed in terms of (traction-like) density functions. Once the density functions are solved, the actual displacements and tractions can be easily computed (Massonet, 1965; Benjumea & Sikarskie, 1972). For some special problems, this method is spectacularly efficient (Banerjee & Driscoll, 1976), because the explicit separation of variables allows one to discard an entire set of equations. One indirect method, which is particularly useful in the rock mechanics context where slip takes place along predefined planes of weakness, is the "displacement discontinuity" method (Crouch & Starfield, 1983). Here, the density functions are "fictitious" displacements. Despite the apparent simplicity of the indirect methods, enthusiasm for them has waned and direct methods are in the ascendancy. In the direct formulation, the integral equations are expressed in terms of the actual physical variables, such as tractions and displacements. Once the boundary unknowns are obtained, then, if desired, the displacements and stresses at selected internal points can be calculated. For brevity, in this book, we deal only with the direct formulation.

1.4.3 BEM in Elasticity

The first direct formulation of the BEM for linear elasticity is usually credited to
Rizzo (1967). Numerical solutions for three-dimensional problems were obtained
by Cruse (1969). More sophisticated algorithms were developed later, borrowing
ideas from the finite element field, notably by Lachat & Watson (1976). Many
workers have revisited the problem of devising efficient schemes to evaluate the
singular integrals that arise in these algorithms. In this context, one might mention the work of Cruse (1974), Guiggiani & Gigante (1990), Huber, Lang, & Kuhn
(1993), Mi & Aliabadi (1996), among others. This area is the subject of an exhaus-
tive review by Sladek & Sladek (1998). The so-called corner problem has also
received a great deal of attention. Here, the difficulty is that, at corners, while
displacements are unique, tractions are multivalued. Depending on the corner
boundary conditions, the integral equations may need to be augmented by additional auxiliary equations (Chaudonneret, 1978; Gao & Davies, 2000a) to close the
equation set. Less satisfactorily, one can "round off" the corner (Jaswon & Symm,
1977), use multiple nodes (Riccardella, 1973), or allow discontinuous elements
(Brebbia & Dominguez, 1992). Using these latter approaches, either resolution
of the corner tractions is diminished or the equation set becomes ill conditioned.

A limitation of the BEM is that it is best suited to those types of problems where
the governing differential equations are known to have an analytical "fundamental" solution. A relevant example is Kelvin's solution (Thomson, 1848), which
yields the displacement field in an infinitely extended linearly elastic isotropic
homogenous solid due to a point force. Based on this fundamental solution, the
BEM permits one to solve problems for finite (and infinite) solids, and indeed
to analyze their elasto–plastic deformations too. However, where a solid is com-
posed of distinct zones of different materials (i.e., a piecewise inhomogeneous
material), a multiregion approach becomes necessary. Examples may be found in
the work of Lachat & Watson (1975), Banerjee & Butterfield (1981), and Kane
et al. (1990). Sometimes, it makes sense to artificially subdivide a single region into
subregions. Although this increases the number of degrees of freedom, it results
in a banded system of equations (rather than a full set) that can be solved more
quickly (Crotty, 1982; Gao & Davies, 2000a). For bodies with high aspect ratios
(i.e., those with high surface-to-volume ratios), subdivision can also improve the
stability of the equation set.

1.4.4 BEM in Elasto–Plasticity

The development of BEM for nonlinear analyses of solids started early (Swedlow
& Cruse, 1971; Riccardella, 1973; Mendelson & Albers, 1975). However, later it
emerged that some subtleties had been overlooked, and corrected formulations
were published by Mukherjee (1977), Bui (1978), and Telles & Brebbia (1979).
Nonlinear analysis requires perforce an incremental approach and until recently
the incremental process has proved to be expensive in computational terms. Also,

accurate evaluation of the strongly singular domain integrals (associated with the interior stresses) has been a stumbling block. To overcome this latter problem, various approaches have been developed. Some of these are inherently inaccurate since they rely on numerical differentiation (Banerjee & Cathie, 1980; Wearing & Dimagiba, 1998). An indirect approach (Telles & Brebbia, 1979; Banerjee, Cathie, & Davies, 1979) requires discretization of the entire region (or subregion). Many direct approaches have been proposed of more or less generality and practical usefulness (e.g., Telles, 1983; Banerjee & Davies, 1984; Banerjee et al., 1989; Gao & Lu, 1992; Guiggiani et al., 1992; Chen, Wang, & Lu 1996; Gao & Davies, 2000b). In Gao & Davies (2000b), a wide-ranging review of such methods may be found. Also, in that paper, a particularly simple and efficient method of performing the singular integration was proposed; this replaces the domain integral by a surface integral that can be readily evaluated by numerical means. To solve the nonlinear system equations, the iterative procedure described by Telles & Brebbia (1979, 1980) based on an initial strain approach is commonly used. However, convergence is slow and implicit solution schemes have been developed more recently (Telles & Carrer, 1991; Bonnet & Mukherjee, 1996). The latter were the first to apply the "consistent tangent operator" approach (Simo & Taylor, 1985) in the BEM context. The alternative "incremental variable stiffness" strategy (Banerjee, Henry & Raveedra, 1989; Banerjee, 1994) eliminates the internal variables and requires no iteration. We advocate a variant (Gao & Davies, 2000b) of this latter approach, in which the plastic multipliers (only) are operated on by a Newton–Raphson iteration scheme; it has proven to be extremely accurate, stable, and efficient. The consequence of these recent significant improvements in nonlinear BEM analysis is that, at last, the method now offers a credible practical alternative to nonlinear finite element analysis.

1.5 Closure

Recent developments in boundary element methods for nonlinear stress analysis in solid mechanics has finally brought the method to the state of maturity demanded by practitioners. In the following chapters, we describe how these theoretical and numerical developments are translated into program code and then demonstrate applications of the code to some practical engineering problems.

CHAPTER TWO

Theory of Elasticity

2.1 Introduction

The purpose of this chapter is to assemble the equations that govern the linear response of solids to loading. These equations form the basis not only for the boundary element code for linear stress analysis but also for nonlinear analysis as well. Of course, there are many excellent texts (e.g., Timoshenko & Goodier, 1970; Fung, 1965) that describe the theory of elasticity in much greater detail and with greater rigor than we can offer here. Suffice it to say that in our brief summary of the theory of elasticity, we will invoke assumptions that, although quite restrictive, are nonetheless reasonably applicable to a wide range of engineering materials.

2.2 Displacements

We assume that solids remain continuous during loading (i.e., no dislocations nor cracks occur). With reference to the right-handed orthogonal set of axes depicted in Fig. 2.1, the displacement **u** may be resolved into three components: u_x, u_y, and u_z or, more conveniently, u_1, u_2, and u_3, which can be represented by the tensor form u_i.

If we further assume that displacements are *sufficiently* small, then the corresponding strains ε_{ij} are defined by the equations

$$\varepsilon_{ij} = (u_{i,j} + u_{j,i})/2 \tag{2.1}$$

where the comma denotes differentiation with respect to the space variable. By way of illustration, consider the case where the subscripts are equal, for example, $i = j = 2 \equiv y$. The above equation yields the familiar result

$$\varepsilon_{yy} = \frac{\partial u_y}{\partial y} \tag{2.2}$$

However, when the subscripts are unequal, for example, $i = 2 \equiv y$, while

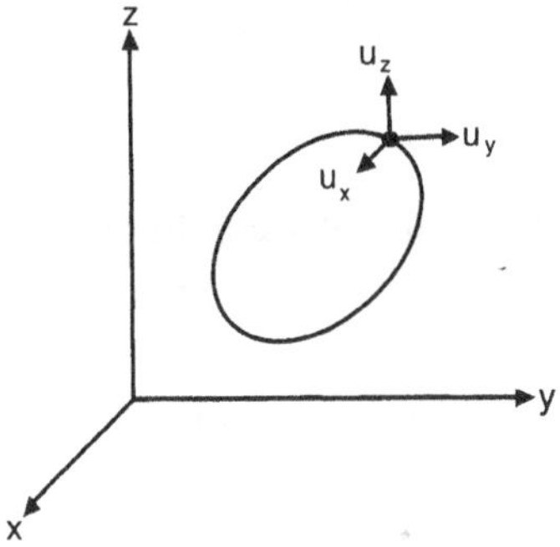

Figure 2.1: Right-handed orthogonal set of axes.

$j = 3 \equiv z$, the equation yields

$$\varepsilon_{yz} = \left(\frac{\partial u_y}{\partial z} + \frac{\partial u_z}{\partial y} \right) \Big/ 2 \qquad (2.3)$$

This result for shear strains may be unfamiliar to those accustomed to the definition of "engineering" shear strains, in which the divisor (2) is absent. These "tensor" shear strains are, by definition, one-half of the magnitude of engineering shear strains. This difference, which arises from the unified tensor equation for strains, creates no particular difficulties provided that the factor of 2 is carried through all subsequent equations.

2.3 Stresses

Stress can be simply defined as "force per unit area." The nine components of the stress tensor σ_{ij} are depicted in Fig. 2.2, with respect to the orthogonal axis system defined earlier. The first superscript defines the face of the element; the second defines the direction of action.

Tensile stresses are, by definition, assumed to be positive. Equilibrium of the elemental cube yields the symmetry conditions for shear stresses, namely,

$$\sigma_{xy} = \sigma_{yx}$$
$$\sigma_{yz} = \sigma_{zy} \qquad (2.4)$$
$$\sigma_{zx} = \sigma_{xz}$$

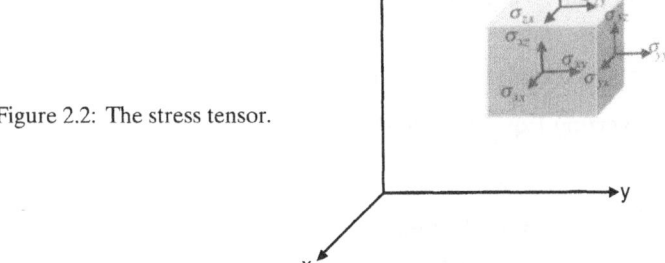

Figure 2.2: The stress tensor.

or, using tensor notation,

$$\sigma_{ij} = \sigma_{ji} \tag{2.5}$$

This symmetry condition, like that for strains, is often invoked without comment in the subsequent theoretical development.

2.4 Stress–Strain Relationships

For an isotropic, linearly elastic material, Hooke's law relates stresses and strains. In algebraic form, this law can be expressed as

$$\varepsilon_{xx} = \frac{1}{E}[\sigma_{xx} - \nu(\sigma_{yy} + \sigma_{zz})]$$

$$\varepsilon_{yy} = \frac{1}{E}[\sigma_{yy} - \nu(\sigma_{zz} + \sigma_{xx})] \tag{2.6}$$

$$\varepsilon_{zz} = \frac{1}{E}[\sigma_{zz} - \nu(\sigma_{xx} + \sigma_{yy})]$$

In these equations, the material constants, E and ν, are the Young's modulus of elasticity and Poisson's ratio, respectively. Tensor shear strains are related to the shear stresses by the shear modulus G, via the equations

$$\varepsilon_{xy} = \sigma_{xy}/2G$$

$$\varepsilon_{yz} = \sigma_{yz}/2G \tag{2.7}$$

$$\varepsilon_{zx} = \sigma_{zx}/2G$$

where the factor of 2 arises as a consequence of our definition of tensor shear strain. Only two of the three material constants (E, G, ν) are independent, since they are linked by the equation

$$G = \frac{E}{2(1+\nu)} \tag{2.8}$$

In some applications of elasticity theory, it is convenient to make use of other material constants, such as the bulk modulus (or modulus of volume expansion)

$$K = \frac{E}{3(1-2\nu)} \tag{2.9}$$

and Lame's constant

$$\lambda = \frac{\nu E}{(1+\nu)(1-2\nu)} \tag{2.10}$$

Inverting Eqs. (2.6) yields

$$\sigma_{xx} = \lambda e + 2G\varepsilon_{xx}$$

$$\sigma_{yy} = \lambda e + 2G\varepsilon_{yy} \tag{2.11}$$

$$\sigma_{zz} = \lambda e + 2G\varepsilon_{zz}$$

where the parameter e denotes the volumetric strain, namely,

$$e = \varepsilon_{xx} + \varepsilon_{yy} + \varepsilon_{zz} \tag{2.12}$$

Because the second term on the right-hand side of Eq. (2.11) is similar to that relating the shear stresses and shear strains, it follows that Hooke's law can be written in the tensor form

$$\sigma_{ij} = \lambda \delta_{ij} \varepsilon_{kk} + 2G \varepsilon_{ij} \tag{2.13}$$

This equation incorporates both the direct and shear terms, by virtue of the properties of the Kronecker delta function (described in Chapter One). More generally, we can rewrite this equation in the form

$$\sigma_{ij} = D^e_{ijkl} \varepsilon_{kl} \tag{2.14}$$

where D^e_{ijkl} is termed the elastic constitutive tensor (or elasticity tensor) defined by the equation

$$D^e_{ijkl} = \lambda \delta_{ij} \delta_{kl} + G(\delta_{ik} \delta_{jl} + \delta_{il} \delta_{jk}) \tag{2.15}$$

The symmetries of the stress and strain tensors are evident in this restatement of the elastic constitutive relationship. Later, when we deal with nonlinear (elasto–plastic) material behavior, we shall see that these elastic constitutive relationships (recast into incremental form) occupy an important role there too. However, it is worth noting that although these (isotropic) stress–strain relationships are adequate for many materials, more complex relationships between stresses and strains are commonly observed in practice, even for elastic (linear) materials. Some of these are briefly discussed in Section 2.7.

2.5 Navier–Cauchy Equations of Equilibrium

If we consider the equilibrium of an infinitesimal cube of material, we can readily derive Cauchy's equations of equilibrium for the stress tensor:

$$\sigma_{ij,i} + b_j = 0 \tag{2.16}$$

where b_j is the body force per unit volume acting in the direction x_j and the comma denotes differentiation with respect to the space variable x_i. Furthermore, on a boundary, with unit outward normal $\mathbf{n}$, the surface tractions must equilibrate the internal stresses, which leads to the equilibrium condition

$$t_j = \sigma_{ij} n_i \tag{2.17}$$

where n_i are the components (direction cosines) of the unit normal vector. It is worth noting that, because of the symmetry of the stress tensor, these two equations are often written in the form $\sigma_{ij,j} + b_i = 0$ and $t_i = \sigma_{ij} n_j$. Now substituting the displacement–strain relationships (Eq. 2.1) and Hooke's law (Eq. 2.13) into Eq. (2.16), we obtain the well-known Navier–Cauchy equations of equilibrium, in terms of displacements:

$$Gu_{i,jj} + (\lambda + G)u_{j,ji} + b_i = 0 \qquad (2.18)$$

(Readers unused to tensor notation may find the discussion of the meaning of this equation in Chapter One helpful.) This equation, together with sufficient boundary conditions, defines the displacement field within a linearly elastic, isotropic, elastic solid. In the boundary element method, the fundamental solution of this equation (the displacement field due to a point force in an infinite solid) is superposed to satisfy these boundary conditions. Here, we obtain a first hint of the power and limitations of the boundary element method: its boundary-only nature and its reliance on the existence of a fundamental solution. These fundamental solutions are central to boundary element methods and we defer further discussion of them to the next chapter.

2.6 Reduced Forms in Two Dimensions

Some problems in solid mechanics can be satisfactorily modeled in two dimensions rather than three. In some case, this simplification arises naturally as a consequence of the physical nature of the problem, but often it is imposed as a convenient approximation to reality. Three distinct cases arise: (a) plane strain, (b) plane stress, and (c) axisymmetry. The first is applicable to cases where the cross section of the solid is sensibly constant over an indefinite (large) distance along one axis; a long embankment (Fig. 2.3) is a typical example in civil engineering. If the loading is also invariant (and imposed "in-plane") along the same axis, then the direct strain in that direction must be zero. By symmetry, the shear strains on the cross-sectional face are also zero; hence, the term plane strain is used.

In contrast, plane stress, as its name suggests, pertains to conditions where the stresses acting on one plane are all zero; an obvious example is a structural plate (Fig. 2.4).

Finally, many engineering structures (e.g., pipes, pressure vessels) are symmetrical about an axis (Fig. 2.5). If the loading system gives rise to a stress system that is independent of the circumferential coordinate θ, then such problems can be modeled by an axisymmetry formulation in the two remaining (r, z) dimensions.

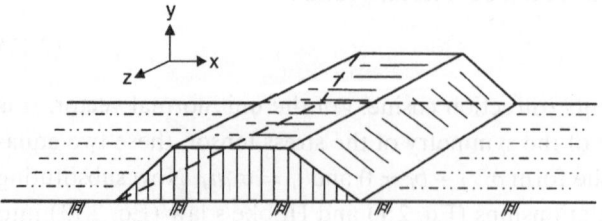

Figure 2.3: Plane strain conditions: an embankment.

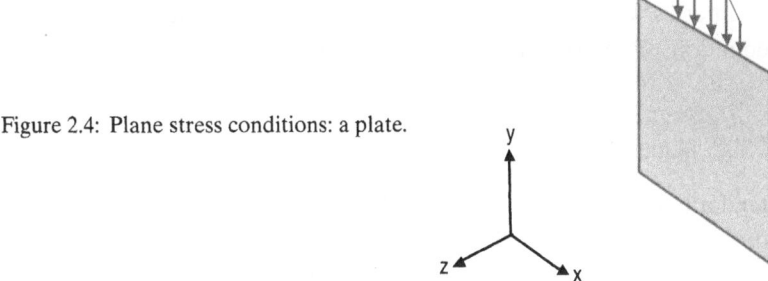

Figure 2.4: Plane stress conditions: a plate.

The reduced sets of equations applicable to each of these three conditions are briefly presented in the following.

2.6.1 Plane Strain

We assume that the cross-sections normal to the third (z) axis are identical and that loading along this axis is invariant. Hence, $u_3 = 0$ and the strains ε_{33}, ε_{32}, and ε_{31} are all zero too. Substituting these values into Eq. (2.6), we immediately obtain the reduced set

$$\sigma_{ij} = \lambda \delta_{ij} \varepsilon_{kk} + 2G\varepsilon_{ij} \tag{2.19}$$

where the indices i, j range from 1 to 2 only. The form of this equation is otherwise identical to the generalized Hooke's law derived earlier. Under elastic conditions, the remaining stress components are

$$\sigma_{33} = \nu(\sigma_{11} + \sigma_{22})$$
$$\sigma_{32} = 0 \tag{2.20}$$
$$\sigma_{31} = 0$$

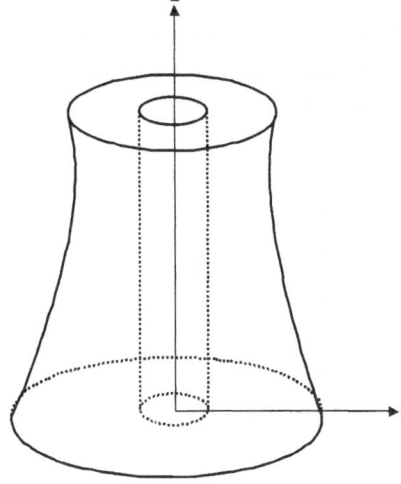

Figure 2.5: Axisymmetric conditions: a thick cylinder.

The governing equations of equilibrium can be readily determined in the same manner as for the three-dimensional case. They differ from the Navier–Cauchy equations (2.18) only in their reduced number of subscripts.

2.6.2 Plane Stress

Under plane stress conditions, we assume that the stresses on the plane normal to the third axis are zero, that is, $\sigma_{33} = \sigma_{32} = \sigma_{31} = 0$. Substituting these values into Eq. (2.6) leads to the constitutive relationships

$$\sigma_{ij} = A\delta_{ij}\varepsilon_{kk} + 2G\varepsilon_{ij} \tag{2.21}$$

where the elastic parameter, A, is equal to $(1 - 2\nu)\lambda/(1 - \nu)$, and the indices i, j, and k range from 1 to 2. The direct strain in the x_3 direction is

$$\varepsilon_{33} = -\nu(\sigma_{11} + \sigma_{22})/E \tag{2.22}$$

whereas the shear strains on this face are both zero. Although plane strain and plane stress conditions are quite different in the physical sense, a simple mathematical equivalence can be established between them. Inspection of the constitutive relationships for plane stress given above reveals that they can be obtained directly from those for plane strain, Eq. (2.19), by simply replacing the Poisson's ratio ν in the plane strain equation by the term $\nu/(1 + \nu)$. This equivalence can be exploited to simplify the computer code for the two cases.

2.6.3 Axisymmetry

It is clearly advantageous to analyze axisymmetrical problems in terms of a cylindrical polar coordinate system (Fig. 2.6), rather than an orthogonal Cartesian system. If the loading as well as the geometry is axially symmetric, then all quantities are invariant with respect to the circumferential angle θ, and displacements may be defined in terms of the radial and axial components (u_r and u_z) only. Thus, these problems can be solved entirely in the r–z plane. Here, the advantages of writing strain–displacement relationships in tensor form are not compelling and

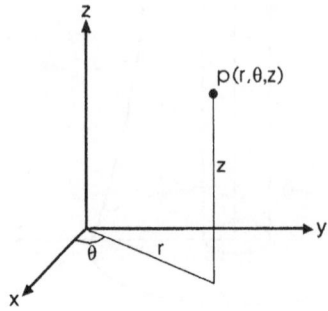

Figure 2.6: Polar cylindrical coordinates.

we revert to the algebraic form:

$$\varepsilon_{rr} = \frac{\partial u_r}{\partial r}$$

$$\varepsilon_{\theta\theta} = \frac{u_r}{r}$$

$$\varepsilon_{zz} = \frac{\partial u_z}{\partial z}$$ \hfill (2.23)

$$\varepsilon_{rz} = \frac{1}{2}\left(\frac{\partial u_r}{\partial z} + \frac{\partial u_z}{\partial r}\right)$$

All other strains are zero. The corresponding stresses ($\sigma_{rr}, \sigma_{\theta\theta}, \sigma_{zz}$, and σ_{rz}) can be obtained directly from the Cartesian form of Hooke's law (Eq. 2.13) by replacing the subscripts r, θ, and z with the Roman subscripts 1, 2, and 3, respectively. If desired, the equations of equilibrium can also be written in terms of cylindrical coordinates which, for axisymmetry, yields the two equations

$$G\left[u_{\alpha,rr} + \frac{1}{r}u_{\alpha,r} + u_{\alpha,zz}\right] + (\lambda + G)\left[\frac{1}{r}(ru_r)_{,r} + u_{z,z}\right]_{,\alpha} + b_\alpha = 0 \qquad (2.24)$$

where the subscript α represents, successively, the radial (r) and axial (z) directions. Whereas the axisymmetry assumption (like plane strain and plane stress) offers the prospect of substantial computational efficiency gains, unfortunately in boundary element methods this expectation is only partially realized. This is because, unlike the two preceding cases, the fundamental solution of the governing differential equations (2.24) cannot be expressed in an explicit analytical form. Given that axisymmetry does not confer any major advantage over the three-dimensional code and that it introduces some additional complexities (of notation, interpretation, etc.), we have not included it in our program. For those interested in developing such code, Becker (1992) has published a Fortran program that deals with the "elastic" case.

2.7 Anisotropic Materials

Many materials, both natural and synthetic, exhibit anisotropic material behavior that evidently cannot be captured by Hookean material models. Given the symmetry of the stress and strain tensors, in the most general case, twenty-one independent material constants are required to characterize anisotropic materials (Lekhnitskii, 1963), as opposed to the two constants in Hooke's law. However, most anisotropic materials display symmetry about one or more axes. In particular, materials are said to be orthotropic if their behavior is symmetric about three orthogonal axes. In this case, at most nine independent constants are required to characterize the material, but this can be reduced to five by invoking a further, but often reasonable, cross-anisotropy (axisymmetric) restriction. Only for this last case does a fundamental solution of the governing equilibrium equations in three dimensions exist in an analytical form (Pan & Chou, 1976) that could reasonably be exploited within a boundary element formulation. For more general cases in three dimensions, it is necessary to construct the fundamental solutions numerically; a laborious process that very few (e.g., Deb, Henry, & Wilson, 1991) have attempted, for obvious reasons. Naturally, in two dimensions, anisotropy proves to be less of an impediment and general methods of attack have been developed to derive fundamental solutions for the general anisotropic case and these have long been exploited within boundary element formulations (Tomlin & Butterfield, 1974; Snyder & Cruse, 1975), with some evidence of success. The increasing use of composites in the manufacturing industry suggests that further refinements in anisotropic analysis would find an eager market. However, this subject goes beyond our core objectives, and we refer readers with a particular interest in it to the references cited above for further information.

2.8 Closure

The review of the principal elements of the theory of elasticity described in this chapter is intended to lay the groundwork for the development of the boundary integral formulation described in the next chapter. There, the governing differential equations of elasticity, which must be satisfied throughout the domain of interest, are transformed into equivalent integral equations, defined over the domain boundary. One consequence of this transformation is immediately apparent: the problem space's dimension reduces by one. Thus, for example, a problem in three-dimensional elasticity reduces to a solution in two-dimensional (surface) space. It might reasonably be expected that this will result in significant computational benefits, as will become apparent later.

CHAPTER THREE

Boundary Integral Equations for Elasticity

3.1 Introduction

In this chapter, we present the boundary integral equations that provide the formal solution to the governing equations of elasticity. These integral equations can be derived in several ways, with various degrees of rigor. Here, we begin by developing the (Betti) reciprocal work theorem using the integration by parts technique described in Chapter One. We then introduce the fundamental solution as one of the two reciprocal elastic states in our derivation of the Somigliana identity. The boundary integral equations then follow as a special case. In addition to these boundary integral equations, we also derive results for internal stresses that will be particularly useful in the subsequent extension to nonlinear problems. Early implementations of these boundary integral equations in elasticity can be traced back to Rizzo (1967) in two dimensions and to Cruse (1969) in three.

3.2 The Kelvin Fundamental Solution

William Thomson (Lord Kelvin) derived the fundamental solution of the Navier–Cauchy equations of equilibrium for an infinitely extended three-dimensional elastic solid, in a paper of remarkable brevity (Thomson, 1848). His results yield the Cartesian components of the displacement field $u_j(q)$, due to a unit point force system $e_i(p)$, in the form

$$u_j(q) = U_{ij}(q,p)e_i(p) \tag{3.1}$$

where

$$U_{ij}(q,p) = \frac{A}{r}[B\delta_{ij} + r_{,i}\, r_{,j}\,] \tag{3.2}$$

and

$$A = 1/[16\pi G(1 - \nu)]$$
$$B = 3 - 4\nu$$
$$r_i = x_i(q) - x_i(p) \tag{3.3}$$
$$r^2 = r_i r_i$$

The points denoted by q and p are termed the field point and source point, respectively. It is important to distinguish carefully between terms such as r_i defined above and $r_{,i}$, which means $\partial r/\partial x_i(q)$ here; the latter quantity is equivalent to r_i/r as shown in Chapter One. It is also essential to observe the order of the subscripts carefully in Eq. (3.1), and indeed in all subsequent equations. Thus, in Eq. (3.1), the summation is taken over the first subscript (i) in the function U_{ij}. Had the equation been written in matrix form (i.e., $\{u\} = [U]\{e\}$), a summation over the second (column) subscript would certainly have been assumed. In tensor notation, the choice is arbitrary but, for reasons that will become apparent later, there is some advantage in the way we have defined U_{ij}. Nevertheless, some authors adopt the opposite convention and great care should be exercised when comparing the two formulations. A more detailed examination of Eq. (3.1) makes it clear that the function $U_{ij}(q,p)$ is singular; that is, it tends to infinity as the source and field points approach each other (as $r \to 0$). This behavior might be expected from a physical perspective as, for example, in the case of a stiletto heel. Given the displacement field, the strain field $\varepsilon_{jk}(q)$ is readily determined from the strain–displacement relationship, which yields

$$\varepsilon_{jk}(q) = E_{ijk}(q,p)e_i(p) \tag{3.4}$$

where

$$E_{ijk}(q,p) = \frac{-A}{r^2}[C(\delta_{ik}r_{,j} + \delta_{ij}r_{,k}) - \delta_{jk}r_{,i} + 3r_{,i}\,r_{,j}\,r_{,k}] \tag{3.5}$$

and

$$C = 1 - 2\nu \tag{3.6}$$

The derivation of this equation is given in Appendix A. Now, making use of the generalized Hooke's law, we can obtain the corresponding stresses σ_{jk} at the field point; thus,

$$\sigma_{jk}(q) = \Sigma_{ijk}(q,p)\,e_i(p) \tag{3.7}$$

where

$$\Sigma_{ijk}(q,p) = \frac{-2GA}{r^2}[C(\delta_{ik}r_{,j} + \delta_{ij}r_{,k} - \delta_{jk}r_{,i}) + 3r_{,i}\,r_{,j}\,r_{,k}] \tag{3.8}$$

Finally, we also need the tractions $t_j(q)$, with respect to a plane defined by the outward normal $n(q)$ at the field point q:

$$t_j(q) = T_{ij}(q,p)\,e_i(p) \tag{3.9}$$

where

$$T_{ij} = \frac{-2GA}{r^2}[C(n_i r_{,j} - n_j r_{,i}) + (3r_{,i} r_{,j} + C\delta_{ij})n_m r_{,m}] \tag{3.10}$$

Further details of the derivations of all these functions, E_{ijk}, Σ_{ijk}, and T_{ij}, are given in Appendix A. Also, in this appendix are the two-dimensional equivalents, for plane strain and plane stress.

3.3 Betti's Reciprocal Work Theorem

To develop the boundary integral equations, we first derive the classical integral identity, known as Betti's reciprocal work theorem, using the method of integration by parts described in Chapter One. We begin by considering two equilibrium states in a region Ω bounded by a surface Γ. The stresses and strains in these two states are denoted by $(\sigma_{ij}, \varepsilon_{ij})$ and $(\sigma_{ij}^*, \varepsilon_{ij}^*)$, respectively.

Using Hooke's law (Eq. 2.13) and multiplying both sides by ε_{ij}^*, we obtain

$$
\begin{aligned}
\sigma_{ij}\varepsilon_{ij}^* &= \lambda\delta_{ij}\varepsilon_{kk}\varepsilon_{ij}^* + 2G\varepsilon_{ij}\varepsilon_{ij}^* \\
&= \lambda\varepsilon_{kk}\varepsilon_{mm}^* + 2G\varepsilon_{ij}\varepsilon_{ij}^* \\
&= (\lambda\delta_{ij}\varepsilon_{mm}^* + 2G\varepsilon_{ij}^*)\varepsilon_{ij} \\
&= \sigma_{ij}^*\varepsilon_{ij}
\end{aligned}
\tag{3.11}
$$

Consequently, the following integral statement holds true:

$$\int_\Omega \sigma_{ij}\varepsilon_{ij}^* d\Omega = \int_\Omega \sigma_{ij}^*\varepsilon_{ij} d\Omega \tag{3.12}$$

This result implies that the work done by the stresses of the first system on the strains of the second system is equal to the work done by the stresses of the second system on the strains of the first system. Now we integrate by parts the left-hand side of this equation as follows:

$$
\begin{aligned}
I_L &= \int_\Omega \sigma_{ij}\varepsilon_{ij}^* d\Omega \\
&= \int_\Omega \sigma_{ij}u_{i,j}^* d\Omega \\
&= \int_\Gamma \sigma_{ij}u_i^* n_j d\Gamma - \int_\Omega u_i^* \sigma_{ij,j} d\Omega \\
&= \int_\Gamma t_i u_i^* d\Gamma + \int_\Omega b_i u_i^* d\Omega
\end{aligned}
\tag{3.13}
$$

In the above derivation, we have made use of the strain–displacement relationships (Eq. 2.1), the symmetry of the stress tensor, Gauss's theorem, and the equilibrium equations (Eqs. 2.16–2.17). Carrying out the analogous operation on the right-hand side of Eq. (3.12) yields

$$I_R = \int_\Gamma t_i^* u_i d\Gamma + \int_\Omega b_i^* u_i d\Omega \tag{3.14}$$

Since, by virtue of Eq. (3.12), $I_L = I_R$, we obtain the following reciprocal identity:

$$\int_\Omega b_i u_i^* d\Omega + \int_\Gamma t_i u_i^* d\Gamma = \int_\Omega b_i^* u_i d\Omega + \int_\Gamma t_i^* u_i d\Gamma \tag{3.15}$$

This is Betti's second reciprocal work theorem, which expresses the equality of the reciprocal work done by two equilibrium states through the domain Ω in terms of, primarily, boundary values. This theorem forms the basis of the Somigliana identity that underpins the boundary element method.

3.4 Somigliana Identity

To develop the Somigliana identity, we first rewrite the reciprocal theorem in the more explicit form (taking the opportunity to substitute the subscript j in place of the subscript i throughout)

$$\int_\Gamma t_j^*(Q) u_j(Q) d\Gamma(Q) + \int_\Omega b_j^*(q) u_j(q) d\Omega(q)$$

$$= \int_\Gamma t_j(Q) u_j^*(Q) d\Gamma(Q) + \int_\Omega b_j(q) u_j^*(q) d\Omega(q) \tag{3.16}$$

where q and Q are points in Ω and Γ, respectively. We now assume that the set (u_j, t_j, b_j) is the real one while the asterisked set (u_j^*, t_j^*, b_j^*) corresponds to that produced by a unit force system e_i^* in an infinite domain. From the definitions of the fundamental solutions, Eqs. (3.1) and (3.9), we obtain

$$t_j^*(q) = T_{ij}(q,p)e_i^*(p)$$
$$u_j^*(q) = U_{ij}(q,p)e_i^*(p) \tag{3.17}$$

Substituting these equations into Eq. (3.16) and, for simplicity, assuming that the real body forces are zero, leads directly to

$$\int_\Gamma T_{ij}(Q,p) e_i^*(p) u_j(Q) d\Gamma(Q) + \int_\Omega b_j^*(q) u_j(q) d\Omega(q)$$

$$= \int_\Gamma t_j(Q) U_{ij}(Q,p) e_i^*(p) d\Gamma(Q) \tag{3.18}$$

To complete the development we need to establish a formal mathematical equivalence between the body force vector $b_j^*(q)$ and the point force vector $e_i^*(p)$. This can be established through the following steps:

$$\int_\Omega b_j^*(q) u_j(q) d\Omega(q) = \int_\Omega e_j^*(q) u_j(q) d\Omega(q)$$

$$= \int_\Omega e_j^*(p)\delta(q,p) u_j(q) d\Omega(q) \tag{3.19}$$

$$= \int_\Omega \delta_{ij} e_i^*(p)\delta(q,p) u_j(q) d\Omega(q)$$

where $\delta(q,p)$ is the Dirac delta function and δ_{ij} is the Kronecker delta. On substituting this result into Eq. (3.18) we observe that the unit force vector $e_i^*(p)$ is common to all integrals. Taking each component of the force vector independently, we obtain, after some rearrangement, the result

$$\int_\Omega \delta_{ij}\delta(q,p)u_j(q)d\Omega(q) = \int_\Gamma U_{ij}(Q,p)t_j(Q)d\Gamma(Q)$$
$$- \int_\Gamma T_{ij}(Q,p)u_j(Q)d\Gamma(Q) \qquad (3.20)$$

The left-hand-side of this equation can be further simplified by noting, from the properties of the Dirac delta function and the Kronecker delta function, that

$$\int_\Omega \delta_{ij}\delta(q,p)u_j(q)d\Omega(q) = \int_\Omega \delta(q,p)u_i(q)d\Omega(q)$$
$$= u_i(p) \qquad (3.21)$$

Strictly speaking, this result is only valid if p is interior to the domain Ω. Obviously, if p is exterior to the domain, then the integral is equal to zero; we defer consideration of the special case when p lies on the boundary of the domain for the moment. Substituting the above result, Eq. (3.21), into Eq. (3.20), we finally obtain the integral equation

$$u_i(p) = \int_\Gamma U_{ij}(Q,p)t_j(Q)d\Gamma(Q) - \int_\Gamma T_{ij}(Q,p)u_j(Q)d\Gamma(Q) \qquad (3.22)$$

Three features of this equation, known as the Somigliana identity (Somigliana, 1885), must be carefully observed:

1. The roles of the subscripts i and j are reversed; the summation is now carried out with respect to the second subscript (j).
2. The integration is carried out with respect to the field point Q and the source point p is now the collocation point.
3. The outward normal n in the kernel function T_{ij} is associated with the surface at the field point Q.

These features are contrary to a superficial common-sense reading of the equations, but they do arise naturally from the reciprocal identities and the kernel function definitions given earlier. Actually, for the kernel function U_{ij}, which is symmetrical with respect to (i, j) and (q, p), these features are immaterial. This is not so for T_{ij}, which is unsymmetrical in both.

The Somigliana identity yields the displacements within the interior of the domain given knowledge of the distribution of both tractions and displacements on the boundary. In a well-posed boundary value problem, exactly one-half of these boundary conditions will be specified and as a consequence the Somigliana identity is insufficient by itself for the purposes of solving such problems. To pursue this path, it is necessary to take the limiting form of the identity as the load point p approaches the boundary, as demonstrated in the following section.

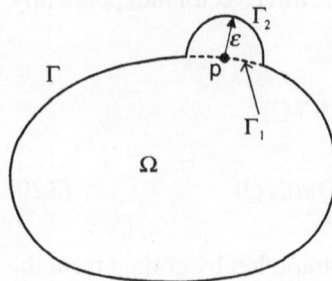

Figure 3.1: Singular point p approaches boundary.

3.5 Boundary Integral Equations

To derive the boundary integral equations, we begin with the Somigliana identity derived in the previous section and determine its limiting form when the load point p approaches the boundary. For the purposes of this demonstration, we assume that the point p is surrounded by part of a spherical surface Γ_2 with radius ε, as depicted in Fig. 3.1.

The Somigliana identity (Eq. 3.22) can be rewritten in the form

$$u_i(p) = \int_{\Gamma - \Gamma_1 + \Gamma_2} U_{ij}(Q,p) t_j(Q) \, d\Gamma(Q) - \int_{\Gamma - \Gamma_1 + \Gamma_2} T_{ij}(Q,p) u_j(Q) \, d\Gamma(Q)$$

(3.23)

where Γ_1 and Γ_2 are the original and auxiliary parts of the boundary at p, respectively. We now consider each of these integrals as $\varepsilon \to 0$ or, equivalently, as $p \to P$. The second integral can be partitioned as follows:

$$\lim_{\varepsilon \to 0} \int_{\Gamma - \Gamma_1 + \Gamma_2} T_{ij}(Q,p) \, u_j(Q) \, d\Gamma(Q) = \lim_{\varepsilon \to 0} \int_{\Gamma_2} T_{ij}(Q,p) \, u_j(Q) \, d\Gamma(Q)$$
$$+ \lim_{\varepsilon \to 0} \int_{\Gamma - \Gamma_1} T_{ij}(Q,p) u_j(Q) d\Gamma(Q)$$

(3.24)

In this equation, the displacement in the first term on the right-hand-side can be isolated, yielding

$$\lim_{\varepsilon \to 0} \int_{\Gamma_2} T_{ij}(Q,p) u_j(Q) d\Gamma(Q) = \lim_{\varepsilon \to 0} \int_{\Gamma_2} T_{ij}(Q,p) [u_j(Q) - u_j(p)] \, d\Gamma(Q)$$
$$+ \lim_{\varepsilon \to 0} \left\{ u_j(p) \int_{\Gamma_2} T_{ij}(Q,p) d\Gamma(Q) \right\} \quad (3.25)$$

As the limit is approached, the first of these integrals evidently vanishes because of displacement continuity ($u_j(Q) = u_j(P)$ in the limit) while the second becomes

$$u_j(P) \left\{ \lim_{\varepsilon \to 0} \int_{\Gamma_2} T_{ij}(Q,p) d\Gamma(Q) \right\} = \beta_{ij}(P) u_j(P)$$

(3.26)

where β_{ij} is a function of the local geometry of the surface at P and may be obtained by analytical integration. Now recalling Eq. (3.24), the second integral on the right hand-side must be interpreted as the so-called Cauchy principal value of the integral. That is, although the integrand is strongly singular at P, the integral is finite if the limits of integration straddle P. This happy result arises from the fact that the integrand is antisymmetrical about P and so this improper integration is perfectly valid. Taking all these results together, in summary we obtain

$$\lim_{\varepsilon \to 0} \int_{\Gamma - \Gamma_1 + \Gamma_2} T_{ij}(Q,p)u_j(Q)\,d\Gamma(Q)$$

$$= \beta_{ij}(P)u_j(P) + \oint_{\Gamma} T_{ij}(Q,P)u_j(Q)d\Gamma(Q) \qquad (3.27)$$

where the notation $\oint$ signifies that the integral is to be interpreted in the Cauchy principal value sense. The same process can be followed for the integral involving the displacement kernel function U_{ij}, but since the order of the singularity is one less than that for the traction kernel function T_{ij}, there are no special difficulties in carrying out this integral: in other words, it can be integrated by normal means. We thus finally arrive at the boundary integral equation

$$c_{ij}u_j(P) = \int_{\Gamma} U_{ij}(Q,P)t_j(Q)d\Gamma(Q) - \oint_{\Gamma} T_{ij}(Q,P)u_j(Q)d\Gamma(Q) \qquad (3.28)$$

where the constant c_{ij} is given by the equation

$$c_{ij} = \delta_{ij} + \beta_{ij}(P) \qquad (3.29)$$

In this equation, the term β_{ij} comes from the limiting process suggested by Eq. (3.26). For a smooth boundary, it turns out that $\beta_{ij} = -\delta_{ij}/2$, but at corners, the algebra becomes more complex and the result depends on the angle subtended by the corner and its orientation in space. However, as will become clearer later, this term need not be calculated analytically, since its value may be determined indirectly by other means.

At this point, following these lengthy analytical manipulations, we should now take stock of where we are. Foremost, we observe that Eq. (3.28) is couched in terms of boundary values (displacements and tractions) only: the boundary integral equations are an exact statement of the relationship between boundary values over a domain governed by elastic material behavior. Clearly, the governing differential equations of equilibrium are also obeyed exactly within the interior of the domain. Thus, the boundary integral equations can, in principle, be solved to obtain an exact solution for any arbitrary domain, given only that sufficient boundary conditions are specified.

3.6 Internal Stresses

The internal stress distribution within an elastic region can be derived directly from the Somigliana identity (Eq. 3.22), given knowledge of the boundary displacements and tractions. First, the strains at the load point are determined from the strain–displacement relationships (Eq. 2.1) as follows:

$$\varepsilon_{ij}(p) = \int_{\Gamma} U_{ijk}^{\varepsilon}(Q,p)t_k(Q)d\Gamma(Q) - \int_{\Gamma} T_{ijk}^{\varepsilon}(Q,p)u_k(Q)d\Gamma(Q) \tag{3.30}$$

where

$$U_{ijk}^{\varepsilon} = (U_{ik,j} + U_{jk,i})/2 \tag{3.31}$$

and

$$T_{ijk}^{\varepsilon} = (T_{ik,j} + T_{jk,i})/2 \tag{3.32}$$

Since the strains are calculated at the load point p from the displacements at this point, the spatial differentiations indicated in these two equations must be carried out with respect to the load point, and not the field point, as has been the case up to this point. Thus, for example, terms such as r,i now become equal to $-r_i/r$; the negative sign arises from the definition of r_i. The corresponding stresses can be determined from Hooke's law and the final result can then be written in the form

$$\sigma_{ij}(p) = \int_{\Gamma} U_{ijk}(Q,p)t_k(Q)d\Gamma(Q) - \int_{\Gamma} T_{ijk}(Q,p)u_k(Q)d\Gamma(Q) \tag{3.33}$$

where

$$U_{ijk} = 2G\left(\frac{\nu}{1-2\nu}\delta_{ij}U_{mmk}^{\varepsilon} + U_{ijk}^{\varepsilon}\right) \tag{3.34}$$

and

$$T_{ijk} = 2G\left(\frac{\nu}{1-2\nu}\delta_{ij}T_{mmk}^{\varepsilon} + T_{ijk}^{\varepsilon}\right) \tag{3.35}$$

Following through these steps, which are similar to those demonstrated in Appendix A in the derivation of the Σ_{ijk} kernel functions, we obtain the explicit expressions

$$U_{ijk} = \frac{1}{8\pi(1-\nu)}\frac{1}{r^2}(C(\delta_{ki}r,_j + \delta_{kj}r,_i - \delta_{ij}r,_k) + 3r,_i\, r,_j\, r,_k) \tag{3.36}$$

and

$$T_{ijk} = \frac{G}{4\pi(1-\nu)}\frac{1}{r^3}\{3r,_m\, n_m[C\delta_{ij}r,_k + \nu(\delta_{ik}r,_j + \delta_{jk}r,_i) - 5r,_i\, r,_j\, r,_k]$$

$$+ 3\nu(n_i r,_j\, r,_k + n_j r,_i\, r,_k) + C(3n_k r,_i\, r,_j + n_j\delta_{ik} + n_i\delta_{jk}) - Dn_k\delta_{ij}\} \tag{3.37}$$

where

$$\begin{aligned} C &= 1 - 2\nu \\ D &= 1 - 4\nu \end{aligned} \tag{3.38}$$

Very similar expressions may be obtained (differing only in some of the constants) for plane strain and plane stress, as shown in Section A.4 of Appendix A. In passing, it may be observed that the kernel function T_{ijk} is hypersingular owing to the presence of the r^3 term in the denominator. Consequently, it can be expected that it will be difficult to calculate stresses accurately near the boundary using Eq. (3.33) and special techniques will be necessary in these cases. Finally, we note that there should be no confusion between the two kernel functions T_{ij} and T_{ijk} as they differ in the number of subscripts.

3.7 Closure

In this chapter, the governing differential equations of elasticity have been transformed into integral equations. An important aspect of this transformation is that the (domain) differential equations reduce to surface integrals, thus reducing the problem-space dimensions by one. Retaining this evident advantage poses difficulties if the formulation is generalized to include consideration of complicating factors such as anisotropy, nonhomogeneities, and body forces. Nevertheless, within fairly restrictive limits, such material, geometric, and loading conditions can be accommodated within a boundary-only integral formulation, as exemplified, for example, by the work of Deb et al. (1991), Gao & Davies (2000a), and Raveendra & Banerjee (1992), although with some cost in complexity and efficiency. However, these extensions go beyond the scope of this study, and here we restrict ourselves to conditions of elastic isotropy in a homogenous single region, free of body forces. It might be observed however that the extension to multiple regions, if their number is not too numerous, is relatively straightforward (Gao & Davies, 2000a). Despite the apparent mathematical complexity of the development of the boundary integral equations in this chapter, the final result (Eq. 3.28) constitutes a remarkably elegant formal solution of the governing equations of elasticity. Unfortunately, except perhaps for some very simple boundary value problems, this integral equation defies solution by analytical means. In practice, numerical techniques (boundary element methods) must be employed to solve it; these techniques are the subject of the next chapter.

CHAPTER FOUR

Numerical Implementation

4.1 Introduction

The boundary integral equations presented in Chapter Three provide an exact description of a solid elastic continuum, and in principle an exact solution can be found for any well-posed set of boundary conditions. In practice, exact closed-form solutions are not possible and recourse must be had to numerical solution techniques. These approximation techniques contain three principal constituents: (a) interpolation, (b) numerical integration (quadrature), and (c) matrix inversion. None of these introduce any fundamental approximations and hence solution accuracy is generally well preserved. Of course, as always, solution accuracy has to be weighed against computational cost, but tolerances of better than 0.1% can be obtained quite routinely. In new applications, convergence studies may be necessary to establish optimal (in terms of the cost–accuracy equation) settings of the various approximation parameters. Only in pathological cases, where special techniques are necessary (e.g., for cracked bodies and slender shell-like structures) are the general techniques described here liable to fail. In such cases, it will be necessary to refer to the specialist literature. As will become evident, if the boundary of the region of interest has corners (or if the region is subdivided into subregions), the boundary integral equations may have to be augmented by auxiliary equations to obtain a closed set. One approach to this problem is described here too. For the most part, in this chapter, we concentrate on the three-dimensional case. The specialization to two dimensions is generally straightforward, but where it is not, we also describe the techniques required for two-dimensional analysis.

4.2 Boundary Discretization

The first step is to subdivide the boundary Γ of the region of interest Ω into a sufficient number (N_e) of elements (e.g., Fig. 4.1). The elements should form a piecewise continuous approximation to the boundary. Naturally, the greater the number of elements employed for this purpose, the better is the approximation, but to preserve computational efficiency it is necessary to specify as few as possible.

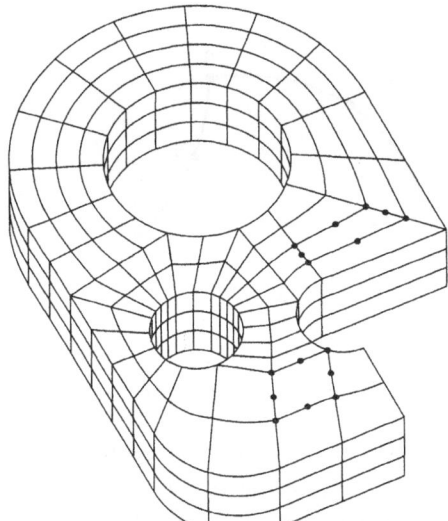

Figure 4.1: Boundary element discretization.

In each element, the global coordinates x_i are interpolated between the coordinates x_i^α of the nodes of that element through interpolation functions, borrowed from the finite element literature (e.g., Zienkiewicz, 1977). Thus,

$$x_i = \sum_{\alpha=1}^{M} N_\alpha(\xi, \eta) x_i^\alpha \qquad (4.1)$$

where M is the number of element nodes and the interpolation functions $N_\alpha(\xi, \eta)$ are commonly referred to as "shape functions." The parameters ξ, η are the local (intrinsic) coordinates, defined by the curvilinear axis system that is everywhere tangential to the element (Fig. 4.2).

By definition, the intrinsic coordinates for an element normally take values in the range ± 1. The shape functions can also be thought of as functions that map the global coordinates of the element into the intrinsic coordinate system. For many purposes, quadratic interpolation functions work very well and these may be defined using the so-called Serendipity eight-noded element. In the intrinsic coordinate system, this element is square (Fig. 4.3) and has three nodes per side.

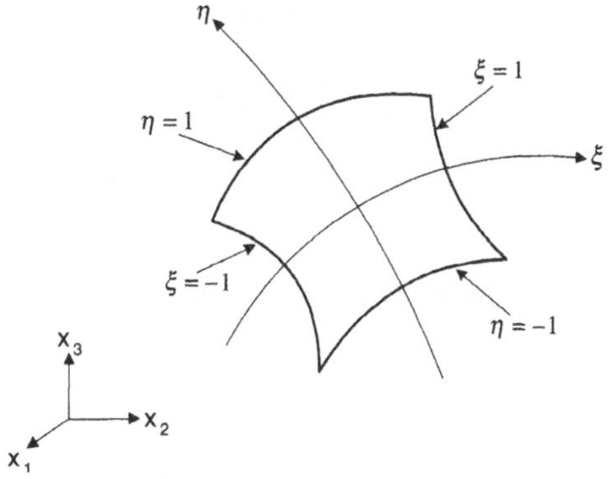

Figure 4.2: Local (intrinsic) coordinate system.

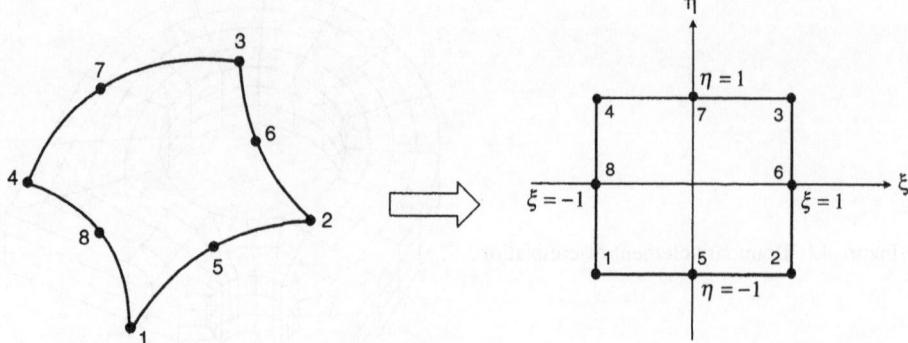

Figure 4.3: The Serendipity eight-noded element.

Although the element is square in the intrinsic coordinate system, the quadratic mapping from the real element permits considerable latitude in the geometry of the latter: In general, the real element may be a curvilinear quadrilateral (as exemplified by Fig. 4.3). However, more extreme distortions are possible and some of these are particularly useful, as will be discussed in Section 4.6. The four corner nodes are associated with the shape functions

$$N_1(\xi, \eta) = \frac{1}{4}(1 - \xi)(1 - \eta)(-1 - \xi - \eta)$$

$$N_2(\xi, \eta) = \frac{1}{4}(1 + \xi)(1 - \eta)(-1 + \xi - \eta)$$

$$N_3(\xi, \eta) = \frac{1}{4}(1 + \xi)(1 + \eta)(-1 + \xi + \eta)$$

$$N_4(\xi, \eta) = \frac{1}{4}(1 - \xi)(1 + \eta)(-1 - \xi + \eta)$$

(4.2)

The pattern of these equations, and their relationship to the intrinsic coordinates of nodes 1–4, should be evident by inspection. Thus, one could rewrite these equations in the form

$$N_c(\xi, \eta) = \frac{1}{4}(1 + \xi\xi_c)(1 + \eta\eta_c)(-1 + \xi\xi_c + \eta\eta_c)$$

(4.3)

where $N_c(\xi, \eta)$ is the shape function for a corner node with intrinsic coordinates ξ_c, η_c. The shape functions of the remaining mid-side nodes are

$$N_5(\xi, \eta) = \frac{1}{2}(1 - \xi^2)(1 - \eta)$$

$$N_6(\xi, \eta) = \frac{1}{2}(1 + \xi)(1 - \eta^2)$$

$$N_7(\xi, \eta) = \frac{1}{2}(1 - \xi^2)(1 + \eta)$$

$$N_8(\xi, \eta) = \frac{1}{2}(1 - \xi)(1 - \eta^2)$$

(4.4)

The quadratic nature of these shape functions is evident in the ξ^2 and the η^2 terms, but they do not constitute a complete polynomial set because there is no

$\xi^2\eta^2$ term. This is not altogether surprising as the full set contains nine terms, whereas there are only eight nodes. This shortcoming can be rectified by introducing a ninth node at the center of the element (the "Lagrangian" element), but it is unclear whether this confers any practical advantage. Naturally, over a single element, the Lagrangian element is superior to the Serendipity element, which means that one can use larger Lagrangian elements for the same degree of accuracy. However, since edge and corner nodes are shared with neighboring elements, the effective number of nodes per element for a Lagrangian element is four (four corner nodes shared by four elements, four mid-side nodes shared by two elements, and one central node) whereas the Serendipity element has only three. Because computational costs are directly linked to the total node count, the advantage of better Lagrangian element accuracy is offset by employing more Serendipity elements. The effectiveness of the quadratic Serendipity element can be demonstrated by examining some specific cases. For example, if a single element is used to model a sphere's surface between the limits of 0 and 90° longitude and 0 and 67° latitude (equator to Arctic Circle), the maximum error in radius is only 3%.

Some special properties of shape functions can be used to verify that they have been encoded correctly. First, each shape function takes the value unity at its "own" node and the value zero at all other nodes. In other words,

$$N_\alpha(\xi_\beta, \eta_\beta) = \delta_{\alpha\beta} \tag{4.5}$$

where the subscript β denotes the value of the intrinsic coordinates at the βth node. This equation implies that the value of the interpolated function (here, the global coordinates x_i) at a node is the nodal value itself; no interpolation is necessary in such cases. Second, it may be verified that the sum of all shape functions, at any arbitrary point, is equal to unity:

$$\sum_{\alpha=1}^{M} N_\alpha(\xi, \eta) = 1 \tag{4.6}$$

The justification for this constraint can be appreciated by considering the special case when all nodal values are equal. Although we have concentrated on the quadratic element here, this is by no means the only option available, nor are such elements restricted to just two intrinsic coordinate systems. Details of some of these elements are described in Appendix B.

4.3 Interpolation of Field Quantities

The variation of displacements and tractions over elements can be described in terms of the element nodal values, in much the same way as the geometry is interpolated between the nodal values. Although not necessary, it is convenient to make use of precisely the same interpolation functions for these field quantities as was done for the geometry. Thus, in such "iso-parametric" formulations, the tractions and displacements, at the intrinsic coordinates (ξ, η), are interpolated between the nodal values, identified by the superscript α, using the equations

$$u_i(\xi, \eta) = \sum_{\alpha=1}^{M} N_\alpha(\xi, \eta) u_i^\alpha$$

$$t_i(\xi, \eta) = \sum_{\alpha=1}^{M} N_\alpha(\xi, \eta) t_i^\alpha \tag{4.7}$$

where α denotes the αth node in the M-noded element. Those familiar with finite element methodology will note that here displacements and tractions need not be interpolated by functions of different order and indeed this would give rise to unnecessary complications later.

4.4 Discretized Boundary Integral Equations

Having defined the displacements and tractions within an element in terms of the element nodal values of displacements and tractions, we can now treat the latter as the discrete variables of the problem. Once these quantities are determined, displacements and tractions on the boundary are known everywhere. Having discretized the boundary into N_e elements (and N nodes), we can now rewrite the boundary integral equations (3.28) in terms of these, yet undetermined, parameters, using the shape functions; thus,

$$c_{ij} u_j(P) = \sum_{e=1}^{N_e} \left\{ \sum_{\alpha=1}^{M} t_j^\alpha(Q) \int_{\Gamma_e} U_{ij}(Q, P) N_\alpha(Q) d\Gamma(Q) \right\}$$
$$- \sum_{e=1}^{N_e} \left\{ \sum_{\alpha=1}^{M} u_j^\alpha(Q) \oint_{\Gamma_e} T_{ij}(Q, P) N_\alpha(Q) d\Gamma(Q) \right\} \tag{4.8}$$

where Γ_e signifies the area of the eth element. To produce a closed set of equations, we choose to write this equation for each node (P) in turn; that is, we collocate at each of the nodes. It may be observed that the integrations of the kernel function–shape function products are carried out over each of the elements. The techniques employed to carry out these integrations are critically important and are discussed in considerable detail in Sections 4.5 and 4.6 of this chapter. For the time being, let us assume that the integrals have been computed with the result that

$$\int_{\Gamma_e} U_{ij}(Q, P) N_\alpha(Q) d\Gamma(Q) = G_{ij}^e(Q, P) \tag{4.9}$$

and

$$\oint_{\Gamma_e} T_{ij}(Q, P) N_\alpha(Q) d\Gamma(Q) = H_{ij}^e(Q, P) \tag{4.10}$$

Thus, from Eq. (4.8), noting that the displacement at P is itself a nodal value, we obtain

$$c_{ij}u_j(P) = \sum_{e=1}^{N_e} \sum_{\alpha=1}^{M} t_j^\alpha(Q) G_{ij}^e(Q, P) - \sum_{e=1}^{N_e} \sum_{\alpha=1}^{M} u_j^\alpha(Q) H_{ij}^e(Q, P) \qquad (4.11)$$

Because the nodal tractions and displacements have yet to be determined, the indicated products and summations must remain as formal statements, which can be better understood in terms of their matrix equivalents, namely,

$$[c]\{u\} = [G]\{t\} - [H']\{u\} \qquad (4.12)$$

where the vectors $\{u\}$ and $\{t\}$ constitute the complete set (of dimension $3N$) of nodal displacements and tractions, respectively, and the matrices $[c]$, $[G]$, and $[H']$ correspond to the coefficients defined earlier. The first of these is, by definition, a "diagonal" matrix (of 3×3 submatrices, in three dimensions). These equations can obviously be condensed into the more convenient form

$$[H]\{u\} = [G]\{t\} \qquad (4.13)$$

where $[H] = [H'] + [c]$. Discussion of methods of solution of this system of equations is deferred to Section 4.11. However, it should be clear that this system is fully populated since, in a single region, there is an interaction between every pair (Q, P) of nodal points. Less obviously, whereas the displacement matrix $[G]$ is symmetric, the traction matrix $[H]$ is not, because the outward normal at the load point P differs, in general, from that at the field point Q.

4.5 Adaptive Integration

Accurate and efficient integration of the displacement and traction integrals over the surface of the problem domain is vitally important. The challenge here is to devise techniques that deliver high accuracy at minimum computational cost. Because the kernel functions must be evaluated many millions of times, significant improvements in run times can be realized by coding the arithmetic efficiently. Beyond that, it is necessary to examine the nature of the kernel functions themselves to ensure that unnecessary function evaluations are eliminated. In general, it is necessary to carry out the integrations numerically, using numerical quadrature methods, and some aspects of these methods are reviewed here for completeness.

In one dimension (say, x), the integral of a function $f(x)$ between arbitrary limits (a, b) can be approximated by the weighted sum of a discrete number of function values, normally evaluated at certain sampling points (ordinates) between the two limits. In general,

$$I = \int_a^b f(x)dx \approx \sum_{k=1}^{n} w_k f(x_k) \qquad (4.14)$$

where x_k and w_k are the kth (of n) ordinates and weights, respectively. For example, in the well-known "midpoint" method, $n = 1$, $x_1 = (a + b)/2$, and $w_1 = b - a$.

Table 4.1. Gaussian quadrature ordinates and
weights (after Stroud & Secrest, 1966)

Order (n)	Ordinates (x_k)	Weights (w_k)
1	0	2
2	±0.57735 02692	1
3	0	0.88888 88889
	±0.77459 66692	0.55555 55556
4	±0.33998 10436	0.65214 51549
	±0.86113 63116	0.34785 48451
5	0	0.56888 88889
	±0.53846 93101	0.47862 86705
	±0.90617 98459	0.23692 68851
6	±0.23861 91861	0.46791 39346
	±0.66120 93865	0.36076 15730
	±0.93246 95142	0.17132 44924

Better accuracy, in general, is obtained but at greater computational cost, by increasing the number of sampling points. If the ordinates are equally spaced on the interval and include the end-points (i.e., $x_1 = a$, $x_n = b$), the resulting quadrature methods constitute the "Newton–Cotes" family. Simpson's rule ($n = 3$) is perhaps the best known of these. However, although the constraint that the ordinates are equally spaced on the interval is convenient in hand calculations, it is quite unnecessary in the machine age. Relaxing this constraint allows us to construct quadrature rules for polynomials that are far more efficient than the Newton–Cotes family. These Gaussian quadrature rules (Stroud & Secrest, 1966) integrate polynomials of order $2n - 1$ exactly, whereas the Newton–Cotes methods integrate (at best) polynomials of order n exactly. Ordinates and weights for Gauss quadrature rules, over the interval ±1 and up to order six, are listed in Table 4.1, to an accuracy of ten significant figures.

To apply the Gauss quadrature rules to an arbitrary interval, it is only necessary to map that interval into Gauss quadrature space, denoted by the symbol ξ, with due consideration for the scaling factor (Jacobian) that this introduces. The mapping from the real interval to Gauss quadrature space yields

$$I = \int_a^b f(x)dx = \int_{-1}^{+1} f(x(\xi))J(x, \xi)d\xi \tag{4.15}$$

and hence, from Eq. (4.14),

$$I \approx \sum_{k=1}^n w_k f(x(\xi))J(x, \xi) \tag{4.16}$$

where $J(x, \xi) = dx/d\xi$. Thus, only in the case of linear mapping can the Jacobian be treated as a constant over the interval. Integration in two and three dimensions can be treated in much the same way, by making use of the so-called product rule.

For example, in two dimensions,

$$I = \int_{a_1}^{b_1} \int_{a_2}^{b_2} f(x_1, x_2) dx_2 dx_1$$

$$= \sum_{k_1=1}^{n_1} \sum_{k_2=1}^{n_2} w_{k_1} w_{k_2} f(x_1(\xi_1, \xi_2), x_2(\xi_1, \xi_2)) J(x, \xi) \qquad (4.17)$$

where

$$J(x, \xi) = \begin{vmatrix} \dfrac{\partial x_1}{\partial \xi_1} & \dfrac{\partial x_1}{\partial \xi_2} \\ \dfrac{\partial x_2}{\partial \xi_1} & \dfrac{\partial x_2}{\partial \xi_2} \end{vmatrix} \qquad (4.18)$$

In general, the limits of integration need not be constants; that is, the physical region need not be defined by boundaries parallel to the global axes x_1 and x_2. This two-dimensional integration might also be employed to integrate a function over a surface in three dimensions if the axes x_1 and x_2 constitute a local (tangential) set. The extension to integration in three dimensions (i.e., over a volume) is straightforward and will not be considered further here, but see Chapter Nine for further details. Although the product rule is convenient, it is not optimal and more efficient strategies have been devised for integrating in n-dimensional space (Stroud & Secrest, 1966). These strategies may be worth revisiting but will not be considered further here.

Returning to our main theme, it should now be apparent that the principal problem is to determine the minimum number of sampling points over a given element consonant with a prescribed accuracy level. This will clearly depend on how close the element is to the nodal collocation point and the properties of the kernel function itself. A complicating factor is that it may be advantageous to distribute the sampling points nonuniformly over the element, perhaps by artificially dividing the element into subregions of integration. This question was first tackled in a rational manner by Lachat & Watson (1976) who made use of certain analytical expressions for the bounds of the error incurred by Gauss integration. For this purpose, the numerical approximation can be replaced by the equation

$$I = \int_a^b f(x) dx = \sum_{k=1}^n w_k f(x_k) + E_n \qquad (4.19)$$

where E_n is the error bound given by

$$E_n = \frac{L^{2n+1}(n!)^4}{(2n+1)[(2n)!]^3} f^{2n}(x) \qquad (4.20)$$

in which $f^{2n}(x)$ signifies the maximum value of the $2n$-th differential of $f(x)$ in the interval and L is the interval width, $b - a$. In the boundary element context, the parameter L can be interpreted as the length of the element along the integration direction. Thus, from Eq. (4.20) we can see that the error of the Gauss quadrature depends on the number of Gauss points and the element size.

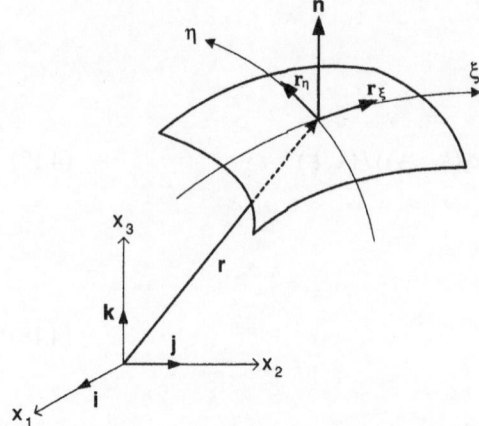

Figure 4.4: Normal vector to a surface element.

To make matters concrete, the integrals that we are concerned with are of the type

$$I_i = \sum_{\alpha=1}^{M} t_j^{\alpha} \int_{-1}^{1} \int_{-1}^{1} U_{ij}[x^P, x(\xi, \eta)] N_\alpha(\xi, \eta) J(\xi, \eta) d\xi d\eta \qquad (4.21)$$

where x^p is the collocation point and U_{ij} is the kernel function to be integrated over the element. We exclude here all singular integrals and hence there is nothing special about the choice of U_{ij} for illustrative purposes rather than T_{ij}. First, we need to determine the Jacobian of the transformation from the global three-dimensional coordinate system to the intrinsic two-dimensional coordinate system of the surface patch. A standard result in differential geometry (e.g., Burke, 1985) is that the Jacobian is equal to the magnitude of the vector cross-product of the vectors $\mathbf{r}_\xi$ and $\mathbf{r}_\eta$, which are directed in the local tangent plane to the surface (Fig. 4.4). In explicit form,

$$J(\xi, \eta) = |\mathbf{r}_\xi \times \mathbf{r}_\eta| \qquad (4.22)$$

where

$$\mathbf{r}_\xi = \frac{\partial x_1(\xi, \eta)}{\partial \xi} \mathbf{i} + \frac{\partial x_2(\xi, \eta)}{\partial \xi} \mathbf{j} + \frac{\partial x_3(\xi, \eta)}{\partial \xi} \mathbf{k}$$

$$\mathbf{r}_\eta = \frac{\partial x_1(\xi, \eta)}{\partial \eta} \mathbf{i} + \frac{\partial x_2(\xi, \eta)}{\partial \eta} \mathbf{j} + \frac{\partial x_3(\xi, \eta)}{\partial \eta} \mathbf{k} \qquad (4.23)$$

in which, $\mathbf{i}, \mathbf{j}$, and $\mathbf{k}$ are the orthogonal unit basis vectors of the global coordinate axes. (Readers unfamiliar with vector operations are referred to Chapter One for a brief explanation of the essentials.) The vector cross-product, by definition, yields a vector $\mathbf{n}^*$ (that is, $\mathbf{n}^* = \mathbf{r}_\xi \times \mathbf{r}_\eta$) that is normal to the surface and so this operation also yields the components, or direction cosines, (n_1, n_2, n_3) of the unit normal vector $\mathbf{n}(= \mathbf{n}^*/|\mathbf{n}^*|)$. These quantities are needed to calculate the kernel function T_{ij} itself. In two dimensions, where integration is carried out over a line element with intrinsic coordinate ξ, the same procedure applies. Here again, the

Jacobian is the magnitude of the normal vector $\mathbf{n}^*$; that is, $J(\xi) = |\mathbf{n}^*|$, where

$$\mathbf{n}^* = \frac{\partial x_2}{\partial \xi}\mathbf{i} - \frac{\partial x_1}{\partial \xi}\mathbf{j} \tag{4.24}$$

Consequently, the components of the unit normal vector $\mathbf{n}$ in this case are simply

$$n_1 = \frac{\partial x_2}{\partial \xi}\Big/ J(\xi)$$

$$n_2 = -\frac{\partial x_1}{\partial \xi}\Big/ J(\xi) \tag{4.25}$$

Following this brief digression, the Gaussian quadrature formula for a surface in three dimensions can be expressed in the intrinsic coordinate system by the equation

$$I = \int_{-1}^{1} \int_{-1}^{1} f(\xi_1, \xi_2) d\xi_1 d\xi_2$$

$$= \sum_{i=1}^{m_1} \sum_{j=1}^{m_2} w_i^1 w_j^2 f(\xi_1^i, \xi_2^j) + E_1 + E_2 \tag{4.26}$$

where (ξ_1^i, ξ_2^j) are the Gauss ordinates, (w_i^1, w_j^2) are the weights, (m_1, m_2) are the Gauss orders, and (E_1, E_2) are the integration errors in the two directions. An approximate formula for the upper bound of the relative error $e_i (= E_i/I)$ in the ith direction has been given by Mustoe (1984):

$$e_i \leq 2\left(\frac{L_i}{4R}\right)^{2m_i} \frac{(2m_i + p - 1)!}{(2m_i)!(p-1)!} \tag{4.27}$$

where p is the order of singularity of the integrand (r^{-p}), L_i is the length of the element in the ith direction, and R is the minimum distance from the source point to the element (as defined in Fig. 4.5).

To avoid using excessively high Gauss orders, elements may be further divided into subelements to reduce the L_i/R ratio. For convenience, Eq. (4.27) may be approximated by the equation (Gao & Davies, 2000c)

$$m_i = \frac{p' \log_e(e_i/2)}{2 \log_e[L_i/(4R)]} \tag{4.28}$$

Figure 4.5: Nonsingular integration.

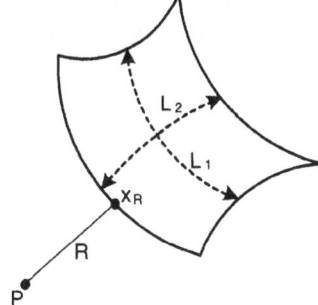

which can be rearranged to yield

$$L_i = 4R\left(\frac{e_i}{2}\right)^{\frac{p'}{2m_i}}$$
(4.29)

where

$$p' = \sqrt{\frac{2}{3}p + \frac{2}{5}}$$
(4.30)

Using this approximation, the required Gauss order is obtained explicitly, rather than through iteration. Alternatively, given a maximal Gauss order, the corresponding subelement dimensions can be obtained explicitly. Although these analytical error bounds are employed widely, a detailed numerical study (Bu & Davies, 1995) of error contours around elements reveals that the contours do not conform very well to the smooth convex shape implied by the error bound. More importantly, in the near field, the error contours differ significantly from the error bounds. Based on this study, Bu & Davies proposed a revised error bound, which may be approximated by the expression

$$m_i = p'[-0.1\log_e(e_i/2)]\left[\left(\frac{8L_i}{3R}\right)^{\frac{3}{4}} + 1\right]$$
(4.31)

where p' is determined as above. Rearranging this equation gives the maximum length L_i of a subelement:

$$L_i = \frac{3}{8}R\left(\frac{-10m_i}{p'\log_e(e_i/2)} - 1\right)^{\frac{4}{3}}$$
(4.32)

Because the analytical bound overestimates error, this revised criterion is generally more efficient. Now, to implement an adaptive integration scheme based on these criteria, it is necessary to devise efficient methods for determining the geometric parameters R and L for each collocation point and for each element or subelement. For convenience here, we denote the complete set of intrinsic coordinates by the notation ξ_i, signifying the two intrinsic axes ξ and η on a surface ($i = 1, 2$), but which might be generalized to include a third axis for volume integration, or simplified to a single axis for a line integration. With this notation, the global Cartesian coordinates at any arbitrary point within an element can be expressed in terms of the nodal coordinates in the form

$$x_j = \sum_{\alpha=1}^{M} N_\alpha(\xi_i)x_j^\alpha$$
(4.33)

where $N_\alpha(\xi_i)$ are the shape functions, M is the number of element nodes and x_j^α are the coordinates of those nodes. The "length" of an element, in the ith intrinsic direction, is defined as the length of the curve through the center of the element (refer to Fig. 4.5). In three dimensions (noting that no summation over the subscript i is intended here) we obtain

$$
L_i = \int_{-1}^{1} \sqrt{\sum_{j=1}^{3} \left(\frac{\partial x_j}{\partial \xi_i} \right)^2} \, d\xi_i
$$

$$
= \int_{-1}^{1} \sqrt{\sum_{j=1}^{3} \left(\sum_{\alpha=1}^{M} \frac{\partial \tilde{N}_\alpha}{\partial \xi_i} x_j^\alpha \right)^2} \, d\xi_i
$$

(4.34)

where $\tilde{N}_\alpha$ is simply the degenerate form of the shape function $N_\alpha(\xi_i)$, in which all intrinsic coordinates, except the ith one, are set to zero; that is, $\tilde{N}_\alpha = N_\alpha(0, \ldots, \xi_i, \ldots, 0)$. To determine the minimum distance (R) from the source point x_j^s to an element is not straightforward, and a Newton–Raphson iterative scheme has been proposed by Gao & Davies (2000c) to accomplish this. This method works well for reasonably well-conditioned discretization schemes but a more robust scheme may be desirable under more difficult circumstances. The essence of the method is to determine the intrinsic coordinates of the point on the element boundary closest to the source point; this point is termed the proximal point for brevity.

First, we begin with some starting guess (ξ_i^0) of the intrinsic coordinates of the source point and we let r_j^0 be the resulting error in the computation of the jth component of the global coordinates of the source point. Now, the notation r_j^k, ξ_i^k is used to denote the values after the kth iteration; that is,

$$
r_j^k = \sum_{\alpha=1}^{M} N_\alpha(\xi_i^k) x_j^\alpha - x_j^s
$$

(4.35)

To obtain improved values of ξ_i, we expand this equation using Taylor's theorem:

$$
r_j^{k+1} = r_j^k + \frac{\partial r_j}{\partial \xi_i} \Delta \xi_i
$$

$$
= r_j^k + \sum_{\alpha=1}^{M} \frac{\partial N_\alpha}{\partial \xi_i} x_j^\alpha \Delta \xi_i
$$

(4.36)

where $\Delta \xi_i$ are the changes in ξ_i. Setting r_j^{k+1} equal to zero, we obtain (in matrix form) the Newton–Raphson iterative scheme

$$
[K^k]\{\Delta \xi\} = -\{r^k\}
$$

(4.37)

where the coefficients of the matrix are

$$
[K_{ji}^k] = \sum_{\alpha=1}^{M} \frac{\partial N_\alpha}{\partial \xi_i} x_j^\alpha
$$

(4.38)

Unless the calculations are being carried out for a volume cell, there will be one less intrinsic coordinate than global coordinate and hence Eq. (4.37) is overprescribed. In that case, the least-squares approximation will suffice; that is,

$$
[K^k]^T[K^k]\{\Delta \xi\} = -[K^k]^T\{r^k\}
$$

(4.39)

where the superscript T denotes the matrix transpose. Solving for $\{\Delta\xi\}$, the current values of ξ_i can be updated and thus

$$\xi_i^{k+1} = \xi_i^k + \Delta\xi_i \tag{4.40}$$

We now make the assumption that the proximal point has the intrinsic coordinates defined by the following equations:

$$\begin{aligned}
\xi_i &= \xi_i^{k+1} && (\text{if}\,|\xi_i^{k+1}| \leq 1) \\
\xi_i &= \mathrm{Sgn}(\xi_i^{k+1}) && (\text{if}\,|\xi_i^{k+1}| \geq 1)
\end{aligned} \tag{4.41}$$

The minimum distance can then be calculated from Pythagoras's theorem as

$$R = \sqrt{\sum_{j=1}^{3} R_j^2} \tag{4.42}$$

where R_j is the jth Cartesian component of the minimum distance and is determined from the intrinsic coordinates of the proximal point:

$$R_j = \sum_{\alpha=1}^{M} N_\alpha(\xi_i)x_j^\alpha - x_j^s \tag{4.43}$$

These calculations (Eqs. 4.35–4.43) are iterated until satisfactory convergence (of R) is attained. However, for highly distorted elements (i.e., elements where the intrinsic coordinate axes are appreciably curved) convergence may be difficult to achieve. In these cases, satisfactory results (within a few percent) have been obtained by simply taking the first (local) minimum value of R. Typically, three iterations are sufficient to attain sufficient convergence, given a starting guess at the element centroid.

Now that the essential components of the adaptive integration are in place, it only remains to devise a sensible strategy to implement it (e.g., Lachat & Watson, 1976; Mustoe, 1984; Dallner & Kuhn, 1993). Here, letting m_{max} be the permitted maximal Gauss order (typically ten), we adopt the following simple but robust strategy:

(a) Calculate element length L_i and minimum distance R to source.
(b) Calculate the Gauss order m_i.
(c) If $m_i \leq m_{max}$ integrate using Gauss quadrature.
(d) If $m_i > m_{max}$, calculate subelement lengths L_i^s, using $m_i = m_{max}$.
(e) Divide element into equal subelements.
(f) Calculate minimum distance R_k from source to kth subelement.
(g) Calculate Gauss order m_i^k for the kth subelement.
(h) Integrate over kth subelement.
(i) Repeat (f)–(h) for all subelements.

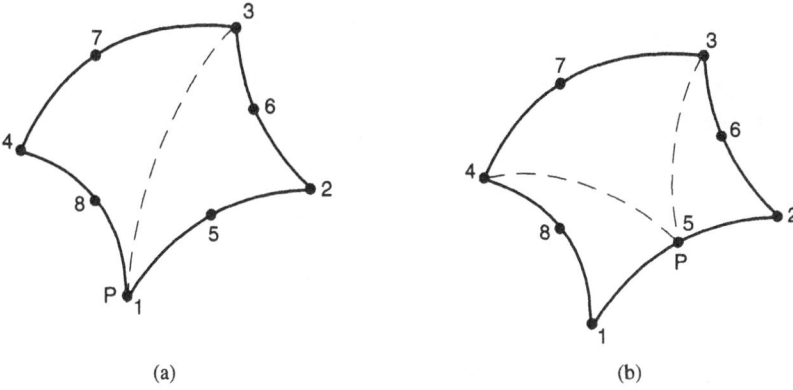

Figure 4.6: Element subdivision for singular integrals.

4.6 Singular Integration

When the load point P is located in the same element as Q, the U_{ij} and T_{ij} kernels become singular because they contain terms of order r^{-1} and r^{-2}, respectively. In this case the direct application of Gaussian quadrature is inadequate and special techniques must be employed to resolve the singularities. These are described next.

4.6.1 Weakly Singular Integrals in Three Dimensions

An element subdivision technique employed by Lachat & Watson (1976) is an effective means for dealing with weakly singular integrals. In this technique (refer to Fig. 4.6), elements containing the source node P are further divided into two triangular subelements (if P is located at a corner node) or three triangular subelements (if P is located at a mid-side node), with P located at the common vertex of these subelements.

The essence of this technique is to map the triangular subelements into the square intrinsic element space (Lachat & Watson, 1976; Mustoe, 1984). Thus, in Fig. 4.7, nodes 1, 8, and 4 of the intrinsic element are all coincident with P. As a result of this degeneracy, it can be shown (Appendix C) that the Jacobian of the transformation is of order r, where r is the distance from the vertex P.

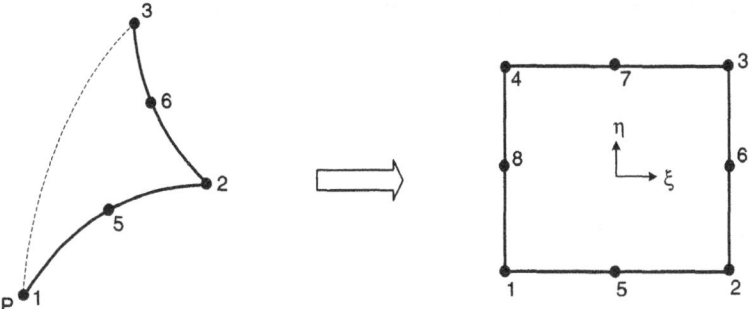

Figure 4.7: Mapping of subelements.

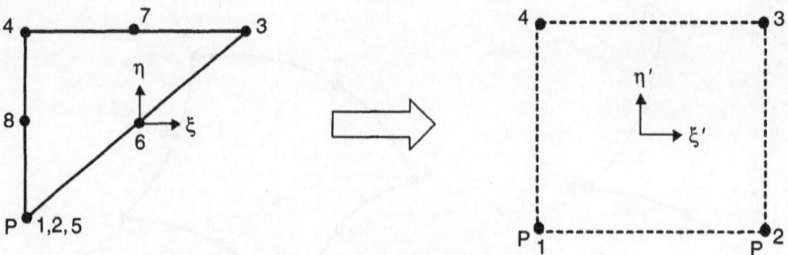

Figure 4.8: Subelement intrinsic coordinate system.

Consequently, this nullifies the weak singularity and the integral can now be evaluated by normal Gauss quadrature.

Because the transformation from element intrinsic coordinates to subelement intrinsic coordinates is linear, an alternative strategy is possible (e.g., Becker, 1992). In this case, a new set of intrinsic coordinates ξ' and η' with their origin at the center of the element is defined for each subelement (refer to Fig. 4.8). Linear shape functions (Fig. 4.9) are now used to determine the original intrinsic coordinates for a point in the new intrinsic coordinate system, as follows:

$$\xi(\xi', \eta') = \sum_{\alpha=1}^{4} N'_\alpha(\xi', \eta')\xi^\alpha$$

$$\eta(\xi', \eta') = \sum_{\alpha=1}^{4} N'_\alpha(\xi', \eta')\eta^\alpha \tag{4.44}$$

where the linear shape functions are

$$N'_1(\xi', \eta') = \frac{1}{4}(1 - \xi')(1 - \eta')$$

$$N'_2(\xi', \eta') = \frac{1}{4}(1 + \xi')(1 - \eta')$$

$$N'_3(\xi', \eta') = \frac{1}{4}(1 + \xi')(1 + \eta') \tag{4.45}$$

$$N'_4(\xi', \eta') = \frac{1}{4}(1 - \xi')(1 + \eta')$$

and the parameters ξ^α and η^α are the nodal values of the original intrinsic coordinates.

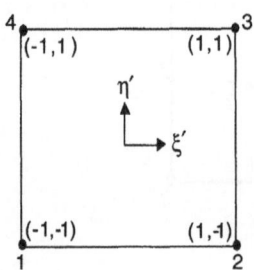

Figure 4.9: Four-noded linear element.

If the element is subdivided into N_s subelements, the singular integrals take the form

$$\int_{\Gamma_e} f(P, Q) d\Gamma = \int_{-1}^{1} \int_{-1}^{1} f(\xi, \eta) J(\xi, \eta) d\xi d\eta$$

$$= \sum_{s=1}^{N_s} \int_{-1}^{1} \int_{-1}^{1} f(\xi'', \eta'') J(\xi'', \eta'') J_s(\xi', \eta') d\xi' d\eta' \qquad (4.46)$$

where ξ'' signifies $\xi(\xi', \eta')$, η'' signifies $\eta(\xi', \eta')$, and $J_s(\xi', \eta')$ is the Jacobian of the transformation from the original to the new intrinsic coordinate system:

$$J_s(\xi', \eta') = \frac{\partial(\xi, \eta)}{\partial(\xi', \eta')}$$

$$= \begin{vmatrix} \dfrac{\partial \xi(\xi', \eta')}{\partial \xi'} & \dfrac{\partial \eta(\xi', \eta')}{\partial \xi'} \\[2ex] \dfrac{\partial \xi(\xi', \eta')}{\partial \eta'} & \dfrac{\partial \eta(\xi', \eta')}{\partial \eta'} \end{vmatrix} \qquad (4.47)$$

As before, the Jacobian $J_s(\xi', \eta')$ tends to zero as $O(r)$ as $r \to 0$, since the original intrinsic coordinates of the two nodes associated with P take the same values in the transformed subelement. For instance, nodes 1 and 2 (in Fig. 4.8) are made coincident by setting

$$\begin{aligned} \xi_1 &= \xi_2 = -1 \\ \eta_1 &= \eta_2 = -1 \end{aligned} \qquad (4.48)$$

In effect, this technique performs the same function as the one described above, and the choice of which to adopt is essentially a matter of personal preference, although it might be argued that the first is a more elegant solution.

4.6.2 Weakly Singular Integrals in Two Dimensions

In two-dimensional space (line integrals), the displacement kernels U_{ij} are weakly singular of order $\log_e(r)$. The strategy we adopt in this case is to isolate the logarithmic singularity and integrate it using the Gauss integration rule for the logarithmically singular functions. Naturally, the nonsingular residual can be easily integrated using ordinary Gauss integration. In the analytical development, we consider the three-noded quadratic element in some detail, while only the final results for the two-noded linear element are given. For the three-noded element, three cases need to be considered because the source P may be located at either one of the end nodes or at the mid-side node. At the first end node (node 1), the distance r between an arbitrary point, defined by its intrinsic coordinate ξ, and the source may be obtained from the equation

$$r^2 = [x(\xi) - x_1]^2 + [y(\xi) - y_1]^2 \qquad (4.49)$$

where x_1 and y_1 are the global coordinates of the source. Substituting the one-dimensional quadratic shape functions (from Appendix B) into this equation, we obtain

$$r^2 = \left[\tfrac{1}{2}(1+\xi)\right]^2\{[-(2-\xi)x_1 + \xi x_2 + 2(1-\xi)x_3]^2$$
$$+ [-(2-\xi)y_1 + \xi y_2 + 2(1-\xi)y_3]^2\} \tag{4.50}$$

Similarly, when P is located at the other end node (node 2), we obtain

$$r^2 = \left[\tfrac{1}{2}(1-\xi)\right]^2\{[-(2+\xi)x_2 - \xi x_1 + 2(1+\xi)x_3]^2$$
$$+ [-(2+\xi)y_2 - \xi y_1 + 2(1+\xi)y_3]^2\} \tag{4.51}$$

These two equations can be expressed in the unified form

$$r^2 = \eta^2\left[f_1^2 + f_2^2\right] \tag{4.52}$$

where

$$\eta = \frac{1}{2}(1 - \xi_p \xi) \tag{4.53}$$

in which $\xi_p = -1$ if the source P is located at the first node and $\xi_p = 1$ if it is located at the other end node (node 2). The functions f_1 and f_2 are then

$$f_1 = -(2 + \xi_p\xi)x_a - \xi_p\xi x_b + 2(1 + \xi_p\xi)x_3$$
$$f_2 = -(2 + \xi_p\xi)y_a - \xi_p\xi y_b + 2(1 + \xi_p\xi)y_3 \tag{4.54}$$

where $a = 1, b = 2$, when the singular point P is at node 1, and $a = 2, b = 1$, when the singular point P is at node 2.

For the case where P is located at the mid-side node (node 3) of the element, we obtain

$$r^2 = \xi^2\left[g_1^2 + g_2^2\right] \tag{4.55}$$

where

$$g_1 = \frac{1}{2}[(\xi - 1)x_1 + (\xi + 1)x_2] - \xi x_3$$
$$g_2 = \frac{1}{2}[(\xi - 1)y_1 + (\xi + 1)y_2] - \xi y_3 \tag{4.56}$$

For the linear element, Eqs. (4.52–4.53) still apply but the functions f_1 and f_2 reduce to the simple expressions

$$f_1 = x_2 - x_1$$
$$f_2 = y_2 - y_1 \tag{4.57}$$

Now, taking the logarithm of Eq. (4.52), we obtain

$$\log_e\left(\frac{1}{r}\right) = \log_e\left(\frac{1}{\eta}\right) - \frac{1}{2}\log_e\left[f_1^2 + f_2^2\right] \tag{4.58}$$

and a similar expression may be obtained from Eq. (4.55). In Eq. (4.58), the second term on the right-hand side is nonsingular and can be integrated without difficulty, while the first term can be integrated using a logarithmically weighted Gauss quadrature rule, as described below. The Gauss quadrature rule for a logarithmic

Table 4.2. Gaussian quadrature ordinates and weights for logarithmically singular functions (after Stroud & Secrest, 1966)

Order (n)	Ordinates (x_k)	Weights (w_k)
1	0.25	1
2	0.11200 88061	0.71853 93190
	0.60227 69081	0.28146 06809
3	0.06389 07931	0.51340 45522
	0.36899 70637	0.39198 00412
	0.76688 03039	0.09461 54066
4	0.04144 84802	0.38346 40681
	0.24527 49143	0.38687 53177
	0.55616 54535	0.19043 51269
	0.84898 23945	0.03922 54871
5	0.02913 44722	0.29789 34717
	0.17397 72133	0.34977 62265
	0.41170 25205	0.23448 82900
	0.67731 41745	0.09893 04595
	0.89477 13610	0.01891 15521

function (Stroud & Secrest, 1966) takes the form

$$I = \int_0^1 \log_e(1/x) f(x) dx \approx \sum_{k-1}^{n} w_k f(x_k) \tag{4.59}$$

where the interval of integration is from zero to unity and the ordinates x_k and the weights w_k are given in Table 4.2. In this instance, the function $f(x)$ is the shape function and two-point integration is generally sufficient.

To apply this method of integration to the mid-side node, the element must be divided into two subelements centered on that node. Because the Gaussian interval of integration is from zero to unity, whereas the range of the intrinsic coordinate system is ± 1, the mapping between these two systems is simply $x = -\xi$ for $\xi < 0$ and $x = +\xi$ for $\xi > 0$, and the Jacobian $J(x, \xi)$ is unity. For the end nodes, the integration can be done over the entire element, using the mapping $x = (1 - \xi\xi_p)/2$ and noting that the Jacobian $J(x, \xi)$ is 2.

4.6.3 Strongly Singular Integrals

The strongly singular integrals in Eq. (4.8), interpreted in the Cauchy principal value sense, together with the constant c_{ij}, yields the diagonal terms of the matrix $[H]$ in Eq. (4.13). Accurate evaluation of these terms is critically important. Although direct evaluation is possible by utilizing certain coordinate transformations (e.g., Guiggiani & Gigante, 1990), an indirect method, which exploits the rigid-body-motion constraint, is far more popular. Three cases relating to (a) finite regions (b) infinite regions, and (c) semi-infinite regions are considered in turn below.

If a finite region is subjected to a unit rigid-body displacement u_j^n in the nth Cartesian direction, the surface tractions must all be zero. In other words, at the kth node, we generate n equations in the form

$$u_j^k = u_j^n$$
$$t_j^k = 0 \tag{4.60}$$

Substituting these equations into Eq. (4.13), we obtain the matrix equations

$$[H]\{I\}^n = \{0\} \tag{4.61}$$

where $\{I\}^n$ is a set of n column vectors, in which (for all nodes) unit displacements are prescribed in the nth direction and zero displacements in all other directions. From Eq. (4.61), the coefficients of the singular submatrix for the kth node (which appears on the leading diagonal of the matrix H) can be determined from the (negative) sum of the off-diagonal elements; therefore,

$$[H]_{ij}^{kk} = (\delta_{km} - 1) \sum_{m=1}^{N} [H]_{ij}^{km} \tag{4.62}$$

where N is the number of nodes, the subscripts i and j range from 1 to 3 in three dimensions, and the superscripts k and m refer to the nodes.

Turning now to infinite region problems, we see that these can generally be understood as the complementary (or exterior) problems to their corresponding finite-region counterparts. Their boundaries can in general be divided into a finite part Γ and an infinite part Γ_∞. Assuming rigid-body displacement and substituting Eq. (4.60) into the boundary integral equation (3.28), we obtain

$$c_{in}(P) + \int_\Gamma T_{in}(P, Q)d\Gamma(Q) + \int_{\Gamma_\infty} T_{in}(P, Q)d\Gamma(Q) = 0 \tag{4.63}$$

The last integral on the left-hand side can be integrated analytically, and so

$$\int_{\Gamma_\infty} T_{in}(P, Q)d\Gamma(Q) = -\delta_{in} \tag{4.64}$$

Consequently, the diagonal submatrix of the traction kernel can be calculated from the equation

$$[H]_{ij}^{kk} = \delta_{ij} + (\delta_{km} - 1) \sum_{m=1}^{N} [H]_{ij}^{km} \tag{4.65}$$

In other words, by comparison with finite-region problems, the coefficients on the leading diagonal of the traction matrix are increased by unity.

For semi-infinite problems, we adopt a very similar approach. The boundary of semi-infinite regions can be divided into a finite part Γ, which includes the half-space surface, and an infinite part, which includes the half-spherical boundary $\Gamma_{H\infty}$ (Fig. 4.10). Again, substituting Eq. (4.60) into the boundary integral

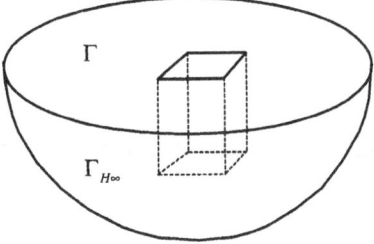

Figure 4.10: Semi-infinite region boundaries.

equation (3.28), we obtain

$$c_{in}(P) + \int_{\Gamma} T_{in}(P, Q)d\Gamma(Q) + \int_{\Gamma_{H\infty}} T_{in}(P, Q)d\Gamma(Q) = 0 \qquad (4.66)$$

The last integral on the left-hand side can be integrated (Gao & Davies, 1998) to give

$$\int_{\Gamma_{H\infty}} T_{in}(P, Q)d\Gamma(Q) = -\frac{1}{2}\delta_{in} \qquad (4.67)$$

Consequently, the diagonal submatrix of the traction kernel can be calculated from the equation

$$[H]_{ij}^{kk} = \frac{1}{2}\delta_{ij} + (\delta_{km} - 1)\sum_{m=1}^{N}[H]_{ij}^{km} \qquad (4.68)$$

For all three cases, the use of the rigid-body constraint obviates the necessity to integrate the strong singularity by rather complex direct methods. Because the indirect method involves only simple arithmetic, then, provided that the integrals of the off-diagonal submatrices are evaluated accurately, it is difficult to justify the use of direct methods.

4.7 Evaluation of Boundary Stresses

The stresses on the boundary of the problem domain cannot be determined directly from the boundary integral equations (Eq. 3.33) because the kernel function T_{ijk} is hypersingular, of order r^{-3}. The most popular technique for overcoming this problem is the so-called traction recovery method (e.g., Cruse, 1974; Telles & Brebbia, 1979; Banerjee & Davies, 1984; Kane, 1994) and, in this section, we describe this approach in explicit terms. The strategy here is to determine the tangential strains (from the displacements) at the point of interest and, hence, using Hooke's law and the known tractions, recover the stresses.

First, we introduce a local Cartesian coordinate system x_i' in which the axes x_1' and x_2' are tangential to the surface and x_3' is directed in the normal direction (Fig. 4.11). The local tangential strains can be expressed in terms of the differentials

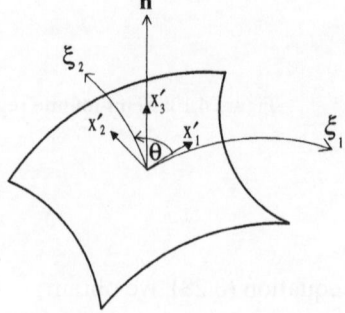

Figure 4.11: Local orthogonal set of axes over a boundary element.

of the displacements as

$$\varepsilon'_{IJ} = \frac{1}{2}\left(\frac{\partial u'_I}{\partial x'_J} + \frac{\partial u'_J}{\partial x'_I}\right) \tag{4.69}$$

where the uppercase subscripts I, J (taking values of 1 and 2) emphasize the fact that these refer to the local axis system. The local derivatives of displacement may be obtained from the equations

$$\frac{\partial u'_I}{\partial x'_J} = \frac{\partial u'_I}{\partial \xi_K}\frac{\partial \xi_K}{\partial x'_J} \tag{4.70}$$

in which the subscript K ranges from 1 to 2 and, for convenience, we adopt an alternative notation (ξ_1, ξ_2) for the intrinsic coordinates in place of (ξ, η). The derivatives of the intrinsic coordinates with respect to the local coordinates are (Lachat, 1975; Becker, 1992)

$$\frac{\partial \xi_1}{\partial x'_1} = \frac{1}{|m_1|}$$

$$\frac{\partial \xi_1}{\partial x'_2} = \frac{-\cos\theta}{|m_1|\sin\theta}$$

$$\frac{\partial \xi_2}{\partial x'_1} = 0 \tag{4.71}$$

$$\frac{\partial \xi_2}{\partial x'_2} = \frac{1}{|m_2|\sin\theta}$$

where θ is the angle defined in Fig. 4.11, and

$$|m_K| = \sqrt{\left(\frac{\partial x_1}{\partial \xi_k}\right)^2 + \left(\frac{\partial x_2}{\partial \xi_k}\right)^2 + \left(\frac{\partial x_3}{\partial \xi_k}\right)^2} \tag{4.72}$$

$$\cos\theta = \frac{1}{|m_1||m_2|}\frac{\partial x_i}{\partial \xi_1}\frac{\partial x_i}{\partial \xi_2} \tag{4.73}$$

The local displacement components in Eq. (4.70) can be expressed in terms of the nodal displacements, referred to the global axis system, via the transformation

$$u'_I = L_{Ij}u_j \tag{4.74}$$

where the displacements (and tractions) can be obtained by interpolation from the nodal values in the usual way, that is,

$$u_i(\xi, \eta) = \sum_{\alpha=1}^{M} N_\alpha(\xi, \eta) u_i^\alpha$$

$$t_i(\xi, \eta) = \sum_{\alpha=1}^{M} N_\alpha(\xi, \eta) t_i^\alpha$$

(4.75)

and L_{Ij} are the direction cosines of the local coordinate system with respect to the global coordinate system. For j in the range 1 to 3, we obtain

$$L_{1j} = \frac{1}{|m_1|} \frac{\partial x_j}{\partial \xi_1}$$

(4.76)

and the remaining terms of this tensor are

$$L_{21} = n_2 L_{13} - n_3 L_{12}$$
$$L_{22} = n_3 L_{11} - n_1 L_{13}$$
$$L_{23} = n_1 L_{12} - n_2 L_{11}$$

(4.77)

where n_1, n_2, and n_3 are the components of the unit normal vector.

Using these equations, we can now calculate the local tangential strains. Then, using Hooke's law and eliminating the local normal strain ε'_{33}, we obtain

$$\sigma'_{11} = \frac{2G}{1-v}(\varepsilon'_{11} + v\varepsilon'_{22}) + \frac{v}{1-v}\sigma'_{33}$$

$$\sigma'_{22} = \frac{2G}{1-v}(\varepsilon'_{22} + v\varepsilon'_{11}) + \frac{v}{1-v}\sigma'_{33}$$

$$\sigma'_{12} = 2G\varepsilon'_{12}$$

(4.78)

Equilibrium requires that

$$\sigma'_{33} = t'_3 = L_{3j}t_j$$
$$\sigma'_{23} = t'_2 = L_{2j}t_j$$
$$\sigma'_{13} = t'_1 = L_{1j}t_j$$

(4.79)

where

$$L_{3j} = n_j$$

(4.80)

Taking these results together, we obtain the following expressions for the stresses, referred to the local set of axes:

$$\sigma'_{11} = \frac{2G}{1-v}\left(\frac{\partial \xi_K}{\partial x'_1}L_{1j} + v\frac{\partial \xi_K}{\partial x'_2}L_{2j}\right)\frac{\partial u_j}{\partial \xi_K} + \frac{v}{1-v}L_{3j}t_j$$

$$\sigma'_{22} = \frac{2G}{1-v}\left(\frac{\partial \xi_K}{\partial x'_2}L_{2j} + v\frac{\partial \xi_K}{\partial x'_1}L_{1j}\right)\frac{\partial u_j}{\partial \xi_K} + \frac{v}{1-v}L_{3j}t_j$$

$$\sigma'_{12} = G\left(\frac{\partial \xi_K}{\partial x'_1}L_{2j} + \frac{\partial \xi_K}{\partial x'_2}L_{1j}\right)\frac{\partial u_j}{\partial \xi_K}$$

(4.81)

Finally, we can employ the transformation relationships

$$\sigma_{mn} = L_{rm} L_{sn} \sigma'_{rs} \tag{4.82}$$

to express the local stresses in terms of their global Cartesian equivalents, as follows:

$$\sigma_{mn} = A_{mnj\alpha} u_j^{\alpha} + B_{mnj} t_j \tag{4.83}$$

After doing some rather lengthy algebra, we obtain the coefficients $A_{mnj\alpha}$ and B_{mnj}:

$$A_{mnj\alpha} = 2G \left\{ \frac{1}{1-\nu} \left[L_{1m} L_{1n} \left(\frac{\partial \xi_K}{\partial x'_1} L_{1j} + \nu \frac{\partial \xi_K}{\partial x'_2} L_{2j} \right) \right. \right.$$
$$\left. + L_{2m} L_{2n} \left(\frac{\partial \xi_K}{\partial x'_2} L_{2j} + \nu \frac{\partial \xi_K}{\partial x'_1} L_{1j} \right) \right]$$
$$\left. + \frac{1}{2} (L_{1m} L_{2n} + L_{2m} L_{1n}) \left(\frac{\partial \xi_K}{\partial x'_1} L_{2j} + \frac{\partial \xi_K}{\partial x'_2} L_{1j} \right) \right\} \frac{\partial N_\alpha}{\partial \xi_K} \tag{4.84}$$

$$B_{mnj} = (L_{3m} L_{1n} + L_{1m} L_{3n}) L_{1j} + (L_{2m} L_{3n} + L_{3m} L_{2n}) L_{2j}$$
$$+ \left(\frac{\nu}{1-\nu} \delta_{mn} + \frac{1-2\nu}{1-\nu} L_{3m} L_{3n} \right) L_{3j} \tag{4.85}$$

The analogous problem arises in two dimensions as well and may be solved in a similar fashion. In this case, the local coordinate ξ is tangential to the element while x'_1 and x'_2 are the local Cartesian axes through the point of interest, as shown in Fig. 4.12.

In the two-dimensional case, the subscripts (m, n, j, α) generally range from 1 to 2. Following the procedure described above, we obtain

$$A_{mnj\alpha} = \frac{2G}{1-\nu} L_{1m} L_{1n} L_{1j} \frac{\partial \xi}{\partial x'_1} \frac{\partial N_\alpha}{\partial \xi} \tag{4.86}$$

$$B_{mnj} = (L_{1m} L_{2n} + L_{2m} L_{1n}) L_{1j} + \left(\delta_{mn} - \frac{1-2\nu}{1-\nu} L_{1m} L_{1n} \right) L_{2j} \tag{4.87}$$

where

$$\frac{\partial \xi}{\partial x'_1} = \frac{1}{J(\xi)} \tag{4.88}$$

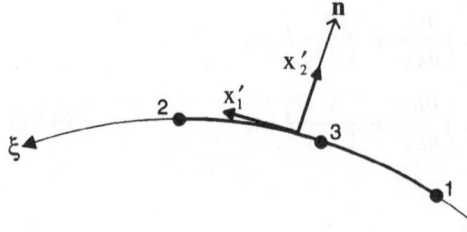

Figure 4.12: Local orthogonal set of axes over a line boundary element.

Here $J(\xi)$ is the Jacobian of the transformation from the global (x_1, x_2) system to the local (ξ) system (refer to Eq. 4.24), and

$$L_{22} = -L_{11} = n_2$$

$$L_{21} = L_{12} = n_1 \tag{4.89}$$

where n_1 and n_2 are the components of the unit normal vector (Eq. 4.25). In plane strain, the above coefficients must be supplemented by the additional terms

$$A_{33j\alpha} = \frac{2Gv}{1-v} L_{1j} \frac{\partial \xi}{\partial x_1'} \frac{\partial N_\alpha}{\partial \xi} \tag{4.90}$$

and

$$B_{33j} = \frac{v}{1-v} L_{2j} \tag{4.91}$$

These equations allow us to compute the stresses on the boundary from the element nodal displacements and tractions. If the stresses are calculated at a node (as is usual) that is shared by several elements, and the stress field is continuous at that point, the results obtained from each of the elements should be averaged. However, if the stress field is discontinuous, because of a traction discontinuity, averaging of all stress components gives erroneous results. In that case, the discontinuity should be preserved by defining multiple nodes at the discontinuity and performing the calculations independently for each element. In nonlinear problems, failure to observe this subtlety can lead to significant error. Finally, although we have expressed the results for boundary stresses in explicit form here (in terms of the coefficients $A_{mnj\alpha}$ and B_{mnj}) it is perfectly satisfactory to perform the various substitutions numerically, if this is preferred.

4.8 Symmetry

Many problems display symmetry about one or more planes. If both the geometry and the boundary conditions are symmetric, then considerable savings in computational time, as well as data preparation effort, can be gained by taking advantage of this fact. In three dimensions, seven symmetry conditions can be identified, as exemplified by Fig. 4.13, if one allows symmetry about any one, any two, or all three Cartesian planes.

Of course, symmetry is not affected by translation (nor rotation) of axes and it may be convenient to define planes of symmetry that are different from the global Cartesian frame. Here, we content ourselves with axis translation only and accordingly define a symmetry plane by its intersection (at the point x_r^c) with the rth Cartesian axis in the global frame. As might be expected, the greater the degree of symmetry, the lesser is the proportion of the domain that must be discretized; for three-fold symmetry this would be only one-eighth of the domain. The coordinates x_i^s of the image (mirrored) nodes, reflected in the plane through

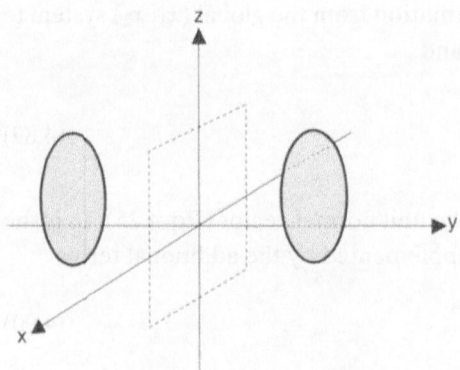

(a) Symmetry about one plane ($y = y^c$)

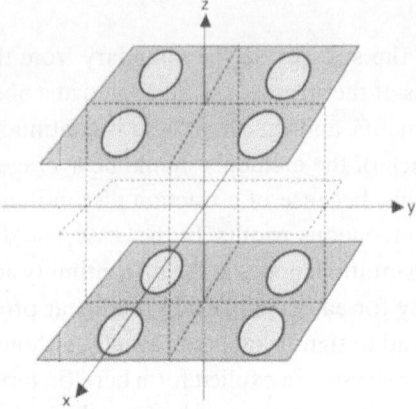

Figure 4.13: Planes of symmetry. Symmetry about (a) one plane and (b) three planes.

(b) Symmetry about three planes ($x = x^c$, $y = y^c$, $z = z^c$)

x_r^c, can be determined from the coordinates x_i^o of the object (original) nodes from the equation

$$x_i^s = x_i^o + 2\delta_{ir}\left(x_r^c - x_r^o\right) \tag{4.92}$$

In image elements (refer to Fig. 4.14), the displacements and tractions in the

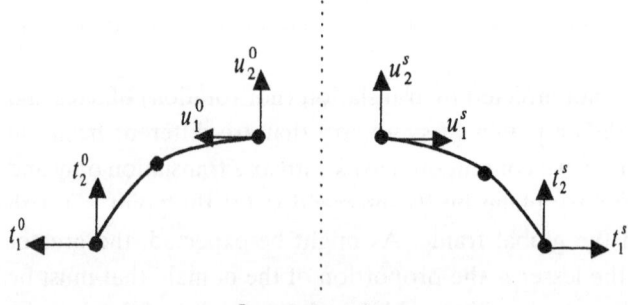

Symmetry plane

Figure 4.14: Traction and displacement symmetry.

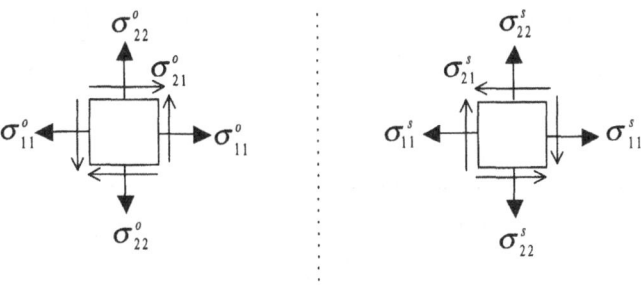

Symmetry plane

Figure 4.15: Stresses in symmetric elements.

direction normal to the plane of symmetry are directed in the opposite sense to those in the original elements, whereas those in the tangential directions are unaltered. Therefore, given symmetry about the plane through x_r^c, the displacements and tractions at the image nodes can be expressed in terms of the values at the original nodes by the equations

$$u_i^s = u_i^o - 2\delta_{ir}u_r^o$$
$$t_i^s = t_i^o - 2\delta_{ir}t_r^o \tag{4.93}$$

The symmetry conditions for stresses are rather more complex, as illustrated in Fig. 4.15 in two dimensions. All stresses in the image element retain the same sign as those of the original element, except for the shear stresses associated with the symmetry plane that cuts through the rth axis. The invariance of the direct stresses (that is, for $i = j$) can be expressed in the form

$$\sigma_{ij}^s = \sigma_{ij}^o \tag{4.94}$$

the relationships for the shear stresses (that is, for $i \neq j$) can be expressed in the form

$$\sigma_{ij}^s = S\sigma_{ij}^o \tag{4.95}$$

where S is equal to -1 if either i or j is equal to r, and S is equal to $+1$ otherwise.

Using Eqs. (4.92)–(4.95), we can express all physical quantities over the image elements in terms of those for the original elements. This does not imply that integrations are no longer necessary over the image elements: integrals over all image elements must still be evaluated, but collocation is necessary only on the original elements. In three-fold symmetry, the number of integrations is reduced by a factor of eight, and the number of system equations is reduced by the same factor. An alternative method of dealing with symmetry, which requires no special coding, is to simply discretize the planes of symmetry using boundary elements and to impose the appropriate boundary conditions on these planes (zero normal displacements and zero tangential tractions). This alternative method will rarely be as computationally efficient as the first method and is always inferior in terms of data preparation and interpretation. Whichever method is adopted, the exploitation of symmetry will almost always result in very significant efficiency gains.

4.9 Corners and Edges

Despite the elegance of the boundary element method, its application to solids that have sharp corners and edges introduces certain practical difficulties. The nub of the problem is that although displacements are uniquely defined at corners, the tractions are multivalued, because each surface has different outward normals. Since the boundary integral equations can yield only one equation per node (per degree of freedom), the number of equations at corner nodes is generally insufficient. Nevertheless, in most practical cases, this difficulty can be easily circumvented by postponing assembly of the system equations (which are generated elementwise), until the boundary conditions are invoked. By this means, the independence of the tractions at any common node is temporarily retained. Then, if only one (or no) traction at a node (per degree of freedom) is unknown, the equations for that node can be assembled incorporating the known boundary conditions. The remaining unknowns at that node will then include no more than one traction (per degree of freedom), which may be obtained by solving the system equations in the normal way. In single-region problems, this simple technique will suffice in all cases, except where displacements are prescribed on the contiguous surfaces at an edge or corner. For these cases, one obvious way to tackle this problem is to "round-off" corners and edges (Jaswon & Symm, 1977), but this is not entirely satisfactory and is impossible in multiregion problems. Alternatively, one could adopt the "unique traction" assumption, namely that the tractions are equal on each contiguous surface, but this is patently false and can give rise to significant errors. Avoiding the problem by using "discontinuous elements" has significant disadvantages in terms of solution stability, computational effort, and accuracy (Wilde, 1998).

To treat corners and edges rationally, it is necessary to introduce additional (coincident) nodes and develop auxiliary equations to determine the additional unknowns (e.g., Chaudonneret, 1978; Zhang & Mukherjee, 1991). If $\tilde{N}$ additional nodes are defined, then, in three dimensions, $3\tilde{N}$ additional (auxiliary) equations must be established. One approach (Gao & Davies, 2000a) begins with the equilibrium equations

$$t_i = \sigma_{ij} n_j \tag{4.96}$$

in which n_j are the components of the unit outward normal. Making use of the local axes system defined earlier (Fig. 4.11), we introduce a local Cartesian coordinate system x_i' ($i = 1, 2, 3$) with the axes x_1' and x_2' tangential to the surface and x_3' in the normal direction. The global quantities, coordinates, and tractions can be transformed into the local coordinate system via the equations

$$x_i' = L_{ij} x_j \tag{4.97}$$

$$t_i' = L_{ij} t_j \tag{4.98}$$

where the primed superscripts identify the local quantities and the transformation tensor is

$$L_{ij} = \frac{\partial x_i'}{\partial x_j} \tag{4.99}$$

From these equations, we immediately obtain

$$\begin{aligned}\frac{\partial t_k'}{\partial x_k'} &= L_{ki} \frac{\partial \sigma_{ij}}{\partial x_k'} n_j \\ &= \frac{\partial \sigma_{ij}}{\partial x_i} n_j\end{aligned} \tag{4.100}$$

From the equilibrium equation,

$$\frac{\partial \sigma_{ij}}{\partial x_i} = 0 \tag{4.101}$$

it follows that

$$\frac{\partial t_k'}{\partial x_k'} = 0 \tag{4.102}$$

This (auxiliary) equation is a general relationship between the tangential components ($k = 1$ and 2 only) of the traction vector and is valid whether the stress field is continuous or not. It might be observed that this equation is simply a restatement of the basic assumptions underlying the equilibrium equation (that the shear stress is constant over an elemental volume) and consequently is valid only over a vanishingly small area. Although derived for three-dimensional problems, it is also valid for two-dimensional problems (with $k = 1$ only). We now need to recast Eq. (4.102) in terms of global quantities and nodal vales of tractions. Using the transformation relations (Eqs. 4.98), we obtain

$$\frac{\partial t_k'}{\partial x_k'} = L_{pi} \frac{\partial t_i}{\partial \xi_k} \frac{\partial \xi_k}{\partial x_p'} \tag{4.103}$$

where $L_{pi}(p = 1, 2)$ are the direction cosines of the local axes, x_1', x_2', respectively (defined in Eq. 4.99). Now, if we make use of the interpolation $t_i = N_\alpha t_i^\alpha$, it follows that

$$\frac{\partial \xi_k}{\partial x_p'} L_{pi} \frac{\partial N_\alpha}{\partial \xi_k} t_i^\alpha = 0 \tag{4.104}$$

where t_i^α is the ith component of the traction at the αth node. This equation can be readily implemented within the boundary element code. Although this approach provides sufficient auxiliary equations in two dimensions, it may be insufficient in some three-dimensional problems. In such cases, we derive supplementary equations by invoking the assumption (which may not necessarily be true) that the stress tensor is continuous at a corner (or edge). For a corner, at the intersection of surfaces S_a and S_b (with unit outward normals n^a and n^b, respectively)

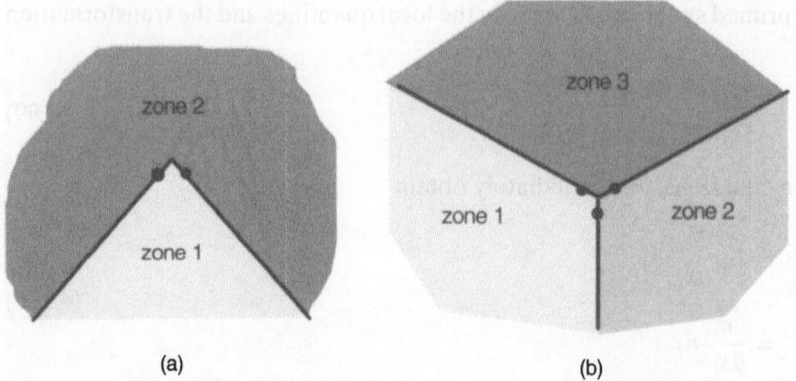

Figure 4.16: Edge intersections (a) between two regions and (b) among three regions.

pre-multiplication of the equilibrium equations by the normal vectors yields

$$n_i^b t_i^a = n_i^b \sigma_{ij} n_j^a$$
$$n_i^a t_i^b = n_i^a \sigma_{ij} n_j^b \qquad (4.105)$$

Because the stress tensor is symmetric, the right-hand sides of these two equations are equal, and hence

$$n_i^b t_i^a = n_i^a t_i^b \qquad (4.106)$$

For single-region problems, Eq. (4.104) yields one equation for each corner node whereas Eq. (4.106) produces one equation for each pair of coincident corner nodes. However, if a node is shared between two regions (in a multiregion problem), the equations can be applied to one region only.

To make matters concrete, consider the case, in three dimensions, of an edge intersection between two regions (Fig. 4.16a). Two coincident nodes need to be defined. After consideration of the interface conditions, there remain nine unknowns (three displacement components and six traction components). Six equations are provided by the boundary integral equations; two auxiliary equations can be obtained from Eq. (4.104) and one from Eq. (4.106). For the more complex three-dimensional edge intersection of three regions (Fig. 4.16b), three coincident nodes need to be defined. After the interface conditions are imposed, twelve unknowns remain. Although the boundary integral equations provide only nine equations, six auxiliary equations – three from Eq. (4.104) and another three from Eq. (4.106) – are available. Experience suggests that the former offers better results because it avoids the restrictive "continuous stress" assumption. Generalizing this discussion, we find that the number of auxiliary equations that must be invoked at corners and edges is

$$m = d \times (n - p + 1) \qquad (4.107)$$

where d is the number of degrees of freedom, n is the number of the nodes associated with the point, and p is the number of the regions meeting at that point.

4.10 Multiple Regions

Multiple-region algorithms have an important role to play in boundary element analysis, and not only to deal with problems in which material properties vary in some piecewise fashion. For example, artificial subdivision of a region with a high surface-to-volume ratio can improve the conditioning of the system equations as well as reduce computational costs. Multiple-region techniques can also be useful in the areas of fracture and nonlinear analysis. Various assembly methods (e.g., Banerjee & Butterfield, 1981; Brebbia, Telles & Wrobel, 1984; Kane et al., 1990) have been proposed but here we follow the substructure technique of Gao & Davies (2000a). The system equations for the ith region can be expressed as

$$[H^i]\{u^i\} = [G^i]\{t^i\} \qquad (4.108)$$

where $\{u^i\}$ and $\{t^i\}$ are the nodal displacements and tractions in the ith region. In general, the traction vector will be supplemented by the additional terms needed to describe the tractions at corners and edges. The nodal terms are now subdivided into two sets: The first set comprises nodal terms solely associated with a single region (termed "external nodes"); the second set comprises those associated with the region interfaces. For convenience, all nodal terms associated with corners and edges are placed in the second set. Equation (4.108) can then be partitioned as follows:

$$\begin{bmatrix} [H_{EE}^i][H_{EI}^i] \\ [H_{IE}^i][H_{II}^i] \end{bmatrix} \left\{ \begin{array}{c} \{u_E^i\} \\ \{u_I^i\} \end{array} \right\} = \begin{bmatrix} [G_{EE}^i][G_{ES}^i] \\ [G_{IE}^i][G_{IS}^i] \end{bmatrix} \left\{ \begin{array}{c} \{t_E^i\} \\ \{t_S^i\} \end{array} \right\} \qquad (4.109)$$

in which the subscript I denotes displacements at the region-to-region interface nodes (Fig. 4.17), the subscript S denotes tractions at the system nodes (interface and additional nodes), and the subscript E denotes the quantities at the remaining external nodes associated with a single region.

The boundary conditions are applied at the regional level and, after the unknowns are shifted to the left-hand side, the block-banded matrix (Eq. 4.108) becomes, after some manipulation,

$$[C_{II}^i]\{u_I^i\} = [D_{IS}^i]\{t_S^i\} + \{y_I^i\} \qquad (4.110)$$

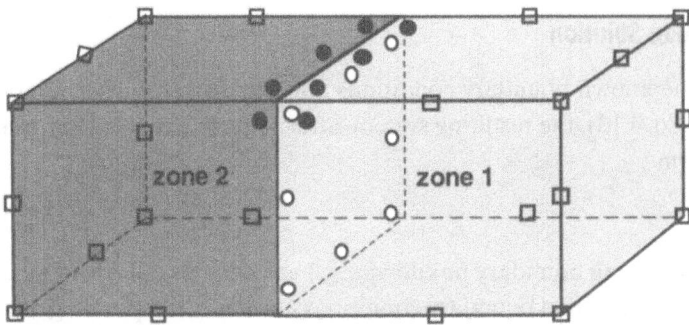

Figure 4.17: Definition of node type in two-region problem (= external node, o = interface node, • = additional node).

It is now convenient to define a global traction vector $\{t_S\}$ for all the interface nodes and all the additional nodes, such that the local traction vector $\{t_S^i\}$ can be expressed in terms of $\{t_S\}$ by the equation

$$\{t_S^i\} = [L^{ti}]\{t_S\} \tag{4.111}$$

where $[L^{ti}]$ is the "traction location matrix" for the ith region, consisting only of zeros and $(\pm)$ unity. The matrix $[L^{ti}]$ expresses the interface equilibrium conditions, for example, the condition $\{t_S^1\} = -\{t_S^2\}$ on the interface between two regions. Similarly, we define a global displacement vector $[u_I]$ for all the interface and additional nodes and an analogous "displacement location matrix" $[L^{ui}]$ for the ith region; thus,

$$\{u_I^i\} = [L^{ui}]\{u_I\} \tag{4.112}$$

where $[L^{ui}]$ identifies the interregional compatibility conditions. Assembling the regional equations, together with the auxiliary equations of the previous section, yields the final system equations:

$$\begin{bmatrix} -[D_{IS}^1][L^{t1}] & [C_{II}^1][L^{u1}] \\ -[D_{IS}^2][L^{t2}] & [C_{II}^2][L^{u2}] \\ \vdots & \vdots \\ -[D_{IS}^m][L^{tm}] & [C_{II}^m][L^{um}] \\ [E_A] & [0] \end{bmatrix} \begin{Bmatrix} \{t_S\} \\ \{u_I\} \end{Bmatrix} = \begin{Bmatrix} \{y_I^1\} \\ \{y_I^2\} \\ \vdots \\ \{y_I^m\} \\ \{y_A\} \end{Bmatrix} \tag{4.113}$$

in which $[E_A]$ and $\{y_A\}$ are generated from the auxiliary equations and the traction boundary conditions. This assembly technique permits simultaneous solution of both displacements and tractions for the interface and additional nodes. However, if a region is not explicitly prevented from undergoing rigid-body displacements, these equations become singular. In this situation, the "singular value decomposition" technique (Press et al., 1992) provides a convenient means of overcoming the problem, albeit at the expense of greater computational time. Numerical problems may also occur if the ratio of the values of the coefficient matrices becomes unduly large. This difficulty can be circumvented by normalizing the coefficients by a representative value of shear modulus.

4.11 System Equation Solution

After substituting the known boundary conditions into the discretized boundary integral equations (Eq. 4.13), the resulting system, after some rearrangement, can be written in the form

$$[A]\{x\} = \{y\} \tag{4.114}$$

where $\{x\}$ are the remaining boundary unknowns, $\{y\}$ are known values obtained from the product of the specified boundary conditions and the corresponding matrix coefficients, and $[A]$ is an array of known coefficients. To avoid numerical problems, associated with poor matrix conditioning, it is desirable that the magnitudes of the tractions and displacements in the vector $\{x\}$ should be broadly equal. One way in which this can be done is to normalize the equations by adopting dimensionless units for tractions and displacements. Thus, beginning with the system equations

$$[H]\{u\} = [G]\{t\} \tag{4.115}$$

we divide the tractions by a representative stress measure (typically, E, Young's modulus of elasticity) and the displacements by a representative length measure (typically the mean element length L). We then obtain

$$[H']\{u'\} = [G']\{t'\} \tag{4.116}$$

where $[G'] = E[G]$ and $[H'] = L[H]$. Because the two coefficient matrices now become of similar magnitude, the conditioning of the final system matrix $[A]$ is much improved. In multiregion problems, a similar procedure can also be adopted although the normalizing factors must be the same for all regions.

Methods for solving these system equations have received relatively little attention in boundary element circles, in contrast to the considerable effort that has been expended in devising efficient solution strategies for finite elements. This difference in emphasis mainly stems from the fact that the number of system equations in boundary element analyses is far fewer and hence their solution constitutes only a small fraction of the total computational effort. Set against this advantage of boundary element methods is that the system equations, for single-region problems, are fully populated and nonsymmetric. The most commonly employed technique is Gaussian matrix reduction and back-substitution, usually with partial (or full) pivoting. If matrix conditioning is improved by adopting the normalizing technique described earlier, then excellent accuracy generally can be obtained by this method even if the rank of the matrix is several hundred or more. In marginal cases, one can always resort to double-precision arithmetic or, better still, the well-known iterative improvement technique (Press et al., 1992):

$$\{x\}^{r+1} = \{x\}^r - \{\delta x\} \tag{4.117}$$

where $\{\delta x\}$ is an estimate of the error in the solution $\{x\}^r$ and is obtained from solving the system equations

$$[A]\{\delta x\} = \{\delta y\} \tag{4.118}$$

and $\{\delta y\}$ is the known error in the current solution, namely,

$$\{\delta y\} = [A]\{x\}^r - \{y\} \tag{4.119}$$

In practice, Eq. (4.118) should be computed using double-precision arithmetic to preserve accuracy, which would otherwise be lost by the subtraction of nearly equal numbers. If this iterative improvement technique is employed, then Gaussian matrix reduction is no longer the method of choice. A matrix decomposition technique (such as the LU method) offers not only comparable solution time for the system equations (4.114) but also re-solution of the iterative equations (4.118) in a fraction of the time. However, for full-scale applications, where the number of degrees of freedom exceed several thousand, equation solution time begins to dominate the solution process and, also, the accuracy of these direct methods may become suspect. In such cases, iterative methods of solution such as the conjugate-gradient method and GMRES methods (Leung & Walker, 1997) offer the prospect of reduced computational time and better accuracy.

4.12 Closure

In this rather lengthy chapter, we have sought to demonstrate the principal techniques that must be employed to translate the formal boundary integral equation solutions of the previous chapter into a practical numerical method. Most of this material will be given concrete expression in Chapter Five, which contains a full description of a computer program that embodies these techniques.

The Elastic Program Code

5.1 Introduction

In this chapter, we describe the Fortran computer program BEMECH, which is based on the preceding theoretical and numerical formulations. This program can deal with plane stress, plane strain, and three-dimensional elastic problems, and it forms the basis for the nonlinear program described in subsequent chapters. The input data for the program is described in Appendix H, while some typical examples are described in Chapter Six.

5.2 Scope of the Program

The program BEMECH is a powerful analysis tool in its own right, capable of analyzing linear single-region problems of arbitrary geometry in both two and three dimensions. Once the boundary unknowns are determined, interior quantities at selected points can be obtained, if desired. In both two and three dimensions, linear and quadratic iso-parametric elements are encoded; in the latter case, quadrilateral (rather than triangular) elements are offered. To reduce computational costs, any one of seven symmetry conditions may be invoked, namely, symmetry about any one, any two, or all three Cartesian planes.

5.3 Program Structure

The program consists of some thirty principal subroutines, which are interlinked as depicted in the flowchart in Fig. 5.1. Subsidiary flowcharts (Figs. 5.2 and 5.3) illuminate the flow paths for the branches emanating from subroutines ADAPTINT and SINGUHG, respectively. The link for the nonlinear analysis is, according to this scheme, quite distinct (see Fig. 5.1). In reality, this separation is not quite so clear cut, but it is nevertheless still easy to handle. In the following, each subroutine is described by (i) a subroutine listing, with a key to theoretical and numerical formulations, and (ii) a glossary of principal local variables with trivial variables omitted. Since these subroutines are arranged in execution order,

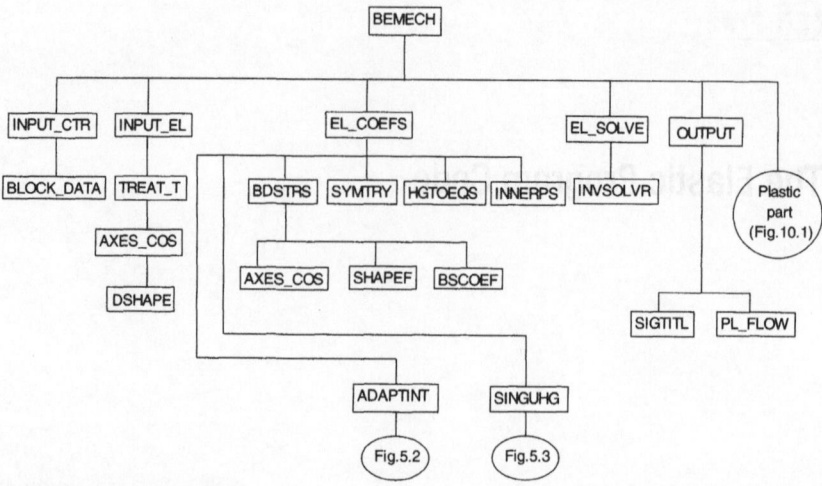

Figure 5.1: Principal flowchart.

an alphabetical index (Table 5.1) has been provided to facilitate the search for specific subroutines. Most subroutines make use of global variables, in addition to local variables, and these are defined in the following section. In addition to their glossary entries here, the input variables are also listed and defined in Appendix H.

5.4 Global Variables

Two types of global variables may be identified. The first of these refers to variables defined in the module program units; the second refers to variables passed as arguments in subroutine calls.

5.4.1 Global Variables in Module Program Units

The two modules VARY_ARRAYS and FIXED_VALUES pass data between program units and precede the other program units. As their names suggest, module VARY_ARRAYS defines allocatable arrays, whereas module FIXED_VALUES defines fixed-size arrays. Some of the global variables are used only for nonlinear analysis: these variables are listed at the end of this section but their definitions are deferred to Chapter Ten.

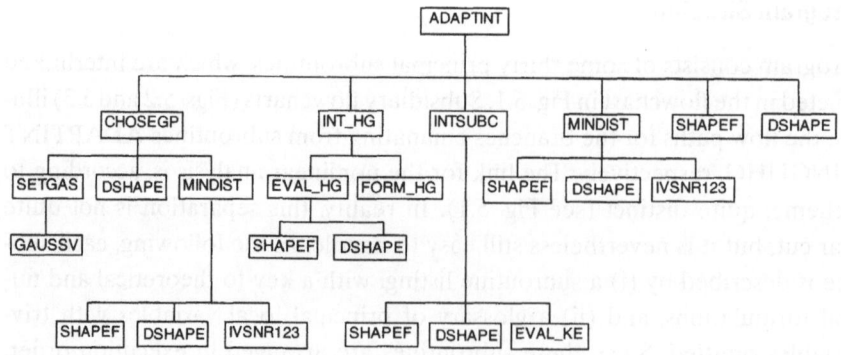

Figure 5.2: Subsidiary flowchart for subroutine ADAPTINT.

Table 5.1. Program units

Name	Section	Description
BEMECH	5.5	Main program
ADAPTINT	5.14	Evaluates nonsingular integrals
AXES_COS	5.10	Direction cosines of local coordinate systems
BDSTRS	5.24	Assembles coefficients in t-recovery method
BSCOEF	5.25	Evaluates coefficients in t-recovery method
BLOCK_DATA	5.7	Initializes variables defined in FIXED_VALUES
CHOSEGP	5.15	Automatically determines Gauss quadrature order
DSHAPE	5.12	Calculates spatial derivatives and Jacobian
DSHAP3D	5.12	Not used – plasticity
EL_COEFS	5.13	Boundary integrals and forms system equations
EL_SOLVE	5.29	Solves system equations for boundary unknowns
EVAL_HG	5.19	Evaluates values of U_{ij} and T_{ij}
FIXED_VALUES	5.4.1	Module: defines global control variables
FORM_HG	5.20	Forms element matrices $[H]$ and $[G]$
GAUSSV	5.16	Sets up Gauss integration points and weights
HGTOEQS	5.27	Assembles matrices $[H]$ and $[G]$ using b.c.
INNERPS	5.28	Forms coefficient matrices for interior points
INPUT_CTR	5.6	Reads execution control data
INPUT_EL	5.8	Reads input data
INT_HG	5.18	Evaluates nonsingular integrals, U_{ij} T_{ij} U_{ijk} T_{ijk}
INVSOLVR	5.30	Calculates matrix inverse and/or solves equations
IVSNR123	5.17	Calculates matrix inverse (rank ≤ 3)
MINDIST	5.17	Calculates minimum distance to an element
OUTPUT	5.31	Output displacements, tractions, and stresses
SETGAS	5.16	Initializes Gauss points and weights
SETDSUB	5.23	Intrinsic coordinates of degenerate subelements
SHAPEF	5.11	Evaluates shape functions
SIGTITL	5.31	Outputs stress component titles
SIN2DHG	5.22	2D singular integrals using log integration
SINGUHG	5.21	Singular integrals (U_{ij}, T_{ij}) using subdivision
SYMTRY	5.26	Deals with symmetry conditions
TREAT_T	5.9	Allocates group tractions to elements
VARY_ARRAYS	5.4.1	Module: defines global allocatable arrays

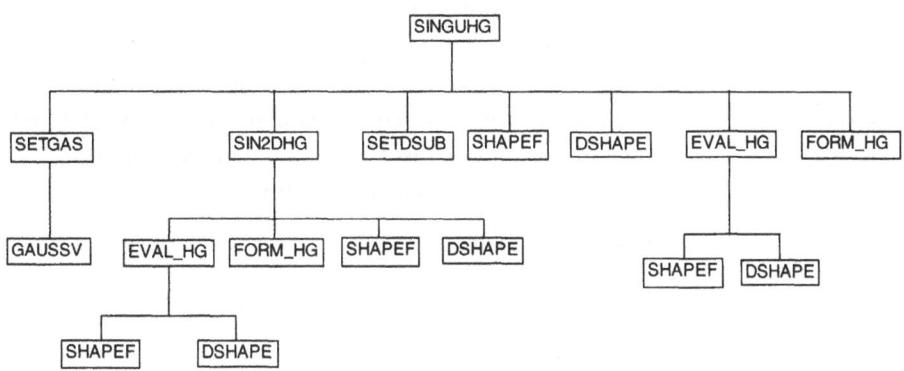

Figure 5.3: Subsidiary flowchart for subroutine SINGUHG.

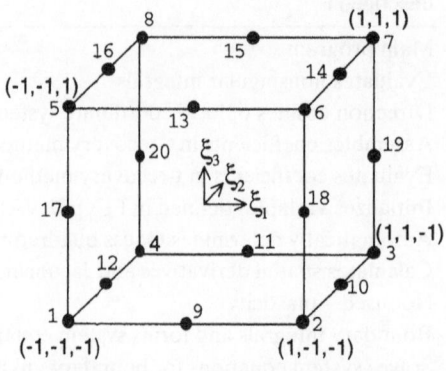

Figure 5.4: Nodal intrinsic coordinates of 20-noded cell.

CD:	Array of nodal coordinates.
CK:	Array of element nodal coordinates.
COEF:	Array containing $[A^u]$, interior displacement coefficients.
CON:	Elastic constant.
CORDL:	Vector of nodal intrinsic coordinates for 20-noded cells (Fig. 5.4). For quadratic elements, nodes 1–4 and 9–12 are used.
COSB:	Vector containing the normal to a boundary element.
CP0:	Zero vector.
CSUB:	Array of intrinsic coordinates of subelement corner nodes.
CSYM:	Vector of the coordinates x_c, y_c, and z_c (cf. KSYM).
DF:	Working space vector.
DIAG:	Array containing the diagonal terms of $[H]$.
DIAGV:	Numerical value $(0, \frac{1}{2}, 1)$ for diagonal term of traction matrix.
DLT:	Array containing the Kronecker delta δ_{ij}.
DN:	Array of $\partial N_\alpha / \partial \xi_k$.
FJCB:	Jacobian $|J|$.
G:	Shear modulus.
GAM:	Elastic constant.
GD:	Array of $\partial x_i / \partial \xi_k$.
GP:	Vector of coordinates of the Gauss points.
GM:	Array containing the matrix $[G]$ for an element.
GW:	Vector of Gauss weighting factors.
HM:	Array containing the matrix $[H]$ for an element.
INX:	Vector that identifies position of stress components in arrays.
KBT:	Vector classifying type of traction boundary condition (cf. KBU).
KBU:	Vector classifying type of displacement boundary condition.
KODP:	Flag identifying type of boundary condition.

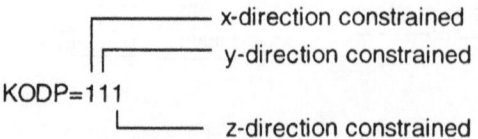

If a flag is switched to zero, no constraint is imposed. KBU takes one of eight values depending on KODP. In three dimensions:

KBU	1	2	3	4	5	6	7	8
KODP	111	000	110	001	011	100	101	010
Specified values	u_x u_y u_z		u_x u_y	u_z	u_y u_z	u_x	u_x u_z	u_y

In two dimensions:

KBU	1	2	3	4
KODP	11	00	10	01
Specified values	u_x u_y		u_x	u_y

KSYM: Symmetry condition flag:

KSYM	0	1	2	3	4	5	6	7
Symmetry about		$x = x_c$	$y = y_c$	$z = z_c$	$x = x_c$ $y = y_c$ $z = z_c$	$y = y_c$ $z = z_c$	$x = x_c$ $z = z_c$	$x = x_c$ $y = y_c$ $z = z_c$

LNDB: Array linking boundary element node and global node number.

LSYM: Vector used to renumber nodes of mirror-image elements, as depicted in Figs. 5.5 and 5.6. For 3-noded line elements, the first two terms are used and the third node is unchanged. For 4-noded quadrilateral elements, the first four are used:

Original node	1	2	3	4	5	6	7	8
Image node	2	1	4	3	5	8	7	6

MSYS: Array that controls system equation assembly, depending on boundary conditions. For three dimensions (eight types of boundary condition) the indices are

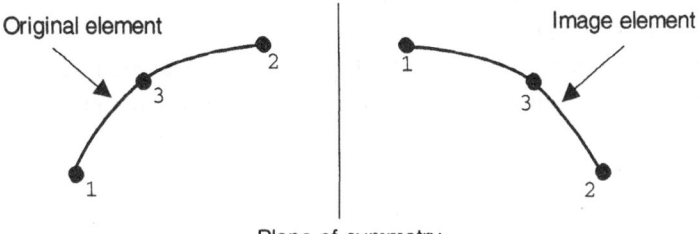

Figure 5.5: Renumbering of line element nodes.

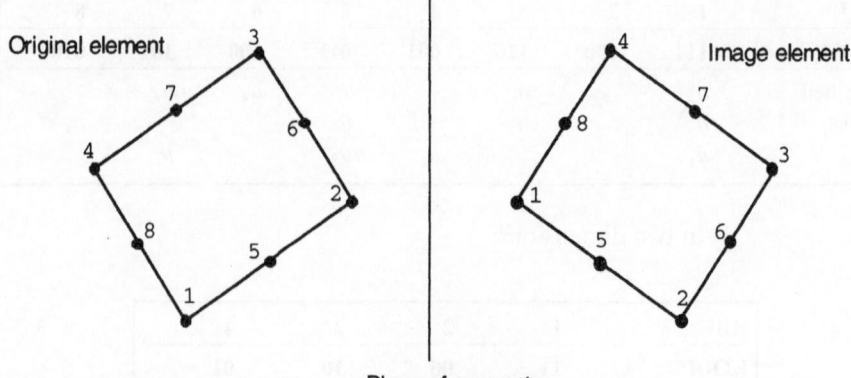

Figure 5.6: Renumbering of quadrilateral element nodes.

1	1	1	3	2	1	1	2
3	3	2	3	3	1	3	2
1	−1	1	1	1	1	2	1

In two dimensions (four types of boundary condition) the indices are

1	1	1	2
2	2	1	2
1	−1	1	1

NAUTO: Gauss order: > 0, all integrations using this Gauss order; = 0, automatically choose Gauss order (≤ 10), but no subdivision; < 0, for singular integrals, Gauss order is |NAUTO|; for nonsingular integrals, automatic Gauss order and subdivision.

NBDM: Dimension of problem boundary, that is, NBDM = NDIM − 1.

NBE: Number of boundary elements.

NBF: Degrees of (displacement) freedom over boundary nodes.

NBP: Total number of boundary nodes.

NBTP: Number of boundary nodes and interior nodes. In linear analyses, NBTP = NTP.

NCELL: Total number of interior cells. Input *zero* in linear analyses.

NCSYM: Number of computational cycles resulting from symmetry:

KSYM	0	1	2	3	4	5	6	7
NCSYM	1	2	2	2	4	4	4	8

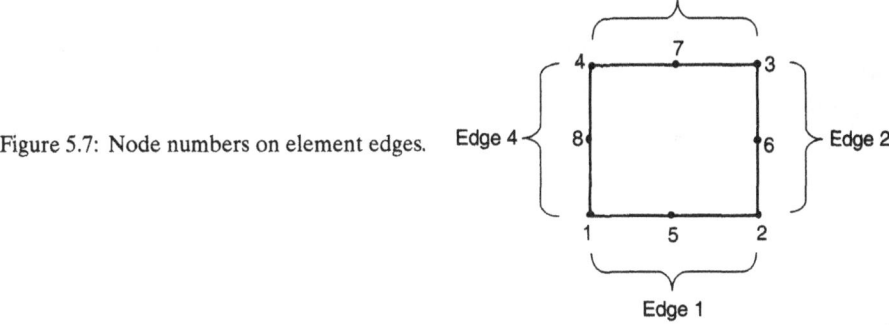

Figure 5.7: Node numbers on element edges.

ND:	Vector of the node numbers of an element or cell.
NDIM:	Dimension of problem: two or three.
NGSIN:	Vector of default Gauss orders along element intrinsic directions for singular integrations. May be changed by the input data.
NGSP:	Vector of Gauss orders in intrinsic directions.
NGSS:	Total number of Gauss integration points over an element.
NNOD:	Vector of numbers of elements meeting at nodes.
NODEF:	Vector of node numbers on each element side (Fig. 5.7).
NSIG:	Number of stress components: 3 – plane stress; 4 – plane strain; 6 – three dimensions.
NSUB:	Vector of numbers of subelements along intrinsic axes.
NSWP:	Vector used to determine the coordinates of an image element from the original element (cf. Fig. 5.6), as tabulated below. NSWP is also used to swap boundary condition types.

Original node	1	2	3	4	5	6	7	8	9	10
Image node	2	1	4	3	6	5	8	7	9	12

Original node	11	12	13	14	15	16	17	18	19	20
Image node	11	10	13	16	15	14	18	17	20	19

NSYM:	Vector that identifies the planes of symmetry defined by KSYM, as tabulated below. In this table, N_{sym} is the number of new elements stemming from each original element.

KSYM	1	2	3	4	5	6	7
N_{sym}	1	1	1	3	3	3	7
NSYM	1	2	3	1 2 1	2 3 2	1 3 1	1 2 1 3 1 2 1

NSYM0:   Initial position of KSYM in array NSYM:

KSYM	1	2	3	4	5	6	7
NSYM0	0	1	2	3	6	9	12

NTF: Number of displacement degrees of freedom.
NTGRP: Vector of group numbers of prescribed tractions. All nodes
              over an element are assumed to have the same traction
              condition.
NTP: Number of boundary and interior nodes.
NTSF: Number of degrees of (stress) freedom over all nodes.
NUGRP: Vector of the group numbers of prescribed displacements.
PI: 3.14159 . . . .
POSGX/Y/Z: Vectors of intrinsic coordinates of Gauss points.
POSGZ: Vector of Gauss coordinates for logarithmic Gauss
              quadrature.
PR: Poisson's ratio.
PR1: $= 1 - \nu$.
PR2: $= 1 - 2\nu$.
PR3: $= 3 - 3\nu$.
PR4: $= 1 - 4\nu$.
PR5: $= 3 - 4\nu$.
PSYM: Vector of symmetry multipliers (see SYMTRY).
RLEVL: Singularity order of integrand.
RT: Array of specified tractions.
RU: Array of specified displacements.
SHAP: Array of element shape functions N_α.
ST: Array containing kernel U_{ijk} for an element.
STRES: Working space vector.
SU: Array containing kernel T_{ijk} for an element.
TOLGP: Tolerance for Gauss integration. Sign identifies criterion
              (> 0, analytical (Eq. 4.28); < 0, numerical (Eq. 4.31)).
WEITX/Y/Z: Vectors of the intrinsic weights for an element.
WEITZ: Vector of Gauss weights for logarithmic quadrature.
X: Vector of unknowns in $[A]\{x\} = \{y\}$.
Y: Vector of known values in $[A]\{x\} = \{y\}$.
YSIG: Vector $\{y^\sigma\}$, internal stresses.

In addition to these global variables, the following are used only for nonlin-
ear analysis (see Chapter Ten): ALFA, BACKS, CEQP, CEQS, CJUP, CLAME,
DFDS, EM, EPSTN, ESTRS, FINCR, FRICT, HARD(I/K/S), INCR, INY, LNDC,
MITER, MULTP, NCRIT, NINCS, NNOD, NODC, NOUT1, NOUT2, PKERN,
RE, ROOT3, TOLER, UNIAX.

```
MODULE VARY_ARRAYS
INTEGER,DIMENSION (:), ALLOCATABLE :: KBU,KBT,NTGRP,NUGRP,NNOD
INTEGER,DIMENSION (:,:), ALLOCATABLE :: LNDB,LNDC
REAL,DIMENSION (:), ALLOCATABLE :: X,Y,YSIG,EPSTN
REAL,DIMENSION (:,:), ALLOCATABLE :: CD,RU,ESTRS,BACKS,DFDS,COEF
END MODULE VARY_ARRAYS


MODULE FIXED_VALUES
INTEGER :: NSIG,NDIM,NBDM,NBP,NBTP,NTP,NBE,NCELL,NBF,NTF,NTSF,      &
& NCSYM,NSYMO,NAUTO,NGSS,NINCS,MITER,NCRIT,NOUT1,NOUT2
REAL :: DIAGV,CON,TOLGP,RLEVL,FJCB,PI,G,PR,PR1,PR2,PR3,PR4,PR5,     &
& CEQS,CEQP,ROOT3,ALFA,CLAME,CJUP,GAM,UNIAX,FRICT,FINCR,TOLER,      &
& HARDI,HARDK,HARDS
DIMENSION DLT(3,3),MSYS(3,8),NODEF(60),CORDL(60),INX(6),INY(9),     &
& CSYM(3),LSYM(8),NSYM(19),NSWP(20),PSYM(6),RE(6),NGSIN(3),NGSP(3), &
& RT(3,8),CK(3,20),GM(3,3,8),HM(3,3,8),SU(6,3,8),ST(6,3,8),NSUB(3), &
& COSB(3),GD(3,3),DN(3,20),DIAG(3,3),ND(20),PKERN(6,120),SHAP(20),  &
& GW(1000),GP(3000),CSUB(3,8),CP0(3),POSGX(10),POSGY(10),POSGZ(10), &
& WEITX(10),WEITY(10),WEITZ(10),STRES(6),DF(6)
END MODULE FIXED_VALUES
```

5.4.2 Global Variables Passed through Argument Lists

Global variables passed through argument lists include:

NCS: Current element singular node number (see Section 5.26).
NDISP: Number of specified displacement groups.
NODE: Number of nodes in boundary element (2, 3, 4, or 8).
NSP: Original singular element node number (see Section 5.26).
NTRAC: Number of specified traction groups.

5.5 Main Program: BEMECH

The main program opens the data files (channels), initializes the dimensions of
the allocatable arrays, and controls the program execution route. The functions of
the channels are as follows:

Channel 5:   Input data file, BEMECH.DAT.
Channel 7:   Output data file, BEMECH.OUT.
Channel 8:   Temporary data file. Stores the specified tractions.
Channel 50:   Temporary data file. Stores the matrix $[A^u]$.
Channels 51–56: Not used – plasticity.

```
PROGRAM BEMECH
USE VARY_ARRAYS; USE FIXED_VALUES
OPEN(5,FILE='BEMECH.DAT',STATUS='OLD')          ! INPUT FILE
```

```
      OPEN(7,FILE='BEMECH.OUT',STATUS='UNKNOWN')      ! OUTPUT FILE
      OPEN(8,STATUS='SCRATCH',ACCESS='DIRECT',RECL=192)
      DO I=50,56; OPEN(I,STATUS='SCRATCH',FORM='UNFORMATTED');ENDDO
!               Input control variables
      CALL INPUT_CTR(NODE,NODC,NDISP,NTRAC)
      ALLOCATE (CD(NDIM,NTP),KBU(NTP),NTGRP(NBE),NUGRP(NTP),KBT(NBE),   &
     &          LNDB(NBE,NODE),LNDC(NODC,NCELL),RU(NDIM,NDISP))
!             Input coordinates,boundary condition et al.
      CALL INPUT_EL(NODE,NDISP,NTRAC)
      ALLOCATE (NNOD(NTP),COEF(NSIG,NTSF),X(NTF),Y(NTF),YSIG(NTSF),      &
     &          EPSTN(NTP),ESTRS(NSIG,NTP),BACKS(NSIG,NTP))
      EPSTN=0.; BACKS=0.   !
!          Compute elastic coefficients and assemble system equations
      CALL EL_COEFS(NODE)
!                   Solve elastic equations
      CALL EL_SOLVE
      WRITE(7,'(//20X," ELASTIC RESULTS :")')
      CALL OUTPUT(NODE,1100,1,1.)
      IF(NCELL.EQ.0) STOP            ! Stop here for elastic problems
!                     Begin plastic part:
      CALL INPUT_NL(NODC)         ! Input cells and plastic properties
      CALL NL_COEFS(NODE,NODC) ! Calculate nonlinear matrices
      DEALLOCATE (KBT,CD,LNDC); ALLOCATE (DFDS(NSIG,NTP))
!             Solve nonlinear system equations by iteration
      CALL NL_SOLVE(NODE,EFSIG)
      WRITE(7,'(//20X," FINAL RESULTS  :")')
      CALL OUTPUT(NODE,NOUT2,1,1.)  ! Print final results
      END PROGRAM BEMECH
```

5.6 Subroutine INPUT_CTR

This routine reads the control variables. In addition to the earlier glossary entries, brief descriptions of these variables, in input order, are given in Appendix H.

```
      SUBROUTINE INPUT_CTR(NODE, NODC,NDISP,NTRAC)
      USE FIXED_VALUES; CHARACTER*6 :: TITLE(20)
      READ(5,'(20A6)')(TITLE(I),I=1,20)              ! Title of problem
      WRITE(7,'(/1X,20A6)')(TITLE(I),I=1,20)
      WRITE(7,'(//," NSIG NODE  NBP  NTP  NBE  NCELL  NTRAC",            &
     &" NDISP  KSYM  NAUTO")')      ! Control variables
      READ(5,*)NSIG,NODE,NBP,NTP,NBE,NCELL,NTRAC,NDISP,KSYM,NAUTO
      WRITE(7,10)NSIG,NODE,NBP,NTP,NBE,NCELL,NTRAC,NDISP,KSYM,NAUTO
      NDIM=NSIG/6+2; NBDM=NDIM-1
      CALL BLOCK_DATA       ! Assign values for commonly used arrays
!          Input Gauss tolerance and symmetry coordinates
      READ(5,*)TOLGP,(CSYM(I),I=1,NDIM); IF(KSYM.EQ.0)                   &
     & WRITE(7,'(/6X,"GP TOLER = ",E13.6,9X,"NO SYMMETRY")')TOLGP
```

```
     IF(KSYM.NE.O) WRITE(7,20)TOLGP,(CSYM(I),I=1,NDIM)
!         Set up other control variables
     NCSYM=2**((KSYM-1)/3+1); IF(KSYM.EQ.0) NCSYM=1
     NSYM0=KSYM+2*IDIM(KSYM-4, 0)-1; NBF=NDIM*NBP; NTSF=NSIG*NTP
     NODC=(1+NODE/3)*4+NODE/8*8    ! Not used -- plasticity
     IF(NAUTO.NE.O) NGSIN=IABS(NAUTO) !Gauss order (singular integrals)
10   FORMAT(I4,I6,I7,2I6,I7,I6,3I7)
20   FORMAT(/6X,'GP TOLER =',E13.6//6X,'SYMMETRIC COORDINATES =',3F14.6)
     END SUBROUTINE INPUT_CTR
```

5.7 Subroutine BLOCK_DATA

This routine assigns values to the variables defined in module FIXED_VALUES.

```
     SUBROUTINE BLOCK_DATA
     USE FIXED_VALUES
!         Set delta function, PI and default Gauss orders
     DLT=RESHAPE((/1.,0.,0., 0.,1.,0., 0.,0.,1./),(/3,3/))
     PI=3.14159265359; NGSIN=(/6,6,6/); CP0=(/0.,0.,0./)
!         Set multipliers for repeated stress components
     IF(NDIM.EQ.2) THEN; RE(1:4)=(/1.,1.,2.,1./)
     INX(1:4)=(/1,3,2,4/); INY(1:4)=(/1,3,3,2/)
     ELSE; RE=(/1.,1.,1.,2.,2.,2./)
     INX=(/1,4,6,2,5,3/); INY-(/1,4,6,4,2,5,6,5,3/); FNDIF
!            Set nodes of each surface of 20-noded cell
     NODEF=(/1, 2, 5,  2,  3,  6,  3,  4,  7,  4,  1,  8,           &
     &        1, 5, 8, 4,17,16,20,12,    1, 2, 6, 5, 9,18,13,17,    &
     &        1, 4, 3, 2,12,11,10, 9,    7, 6, 2, 3,14,18,10,19,    &
     &        7, 3, 4, 8,19,11,20,15,    7, 8, 5, 6,15,16,13,14/)
!            Set nodal intrinsic coordinates for 20-noded cell
     CORDL=(/-1.,-1.,-1.,  1.,-1.,-1.,  1., 1.,-1.,  -1., 1.,-1.,   &
     &         -1.,-1., 1.,  1.,-1., 1.,  1., 1., 1.,  -1., 1., 1.,  &
     &          0.,-1.,-1.,  1., 0.,-1.,  0., 1.,-1.,  -1., 0.,-1.,  &
     &          0.,-1., 1.,  1., 0., 1.,  0., 1., 1.,  -1., 0., 1.,  &
     &         -1.,-1., 0.,  1.,-1., 0.,  1., 1., 0.,  -1., 1., 0./)
!              Set symmetry condition parameters
     LSYM=(/2,1,4,3,5,8,7,6/)
     NSYM=(/1,2,3, 1,2,1, 2,3,2, 1,3,1, 1,2,1,3,1,2,1/)
     NSWP=(/2,1,4,3,6,5,8,7,9,12,11,10,13,16,15,14,18,17,20,19/)
     IF(NDIM.EQ.2) NSWP(5:8)=LSYM(5:8)   ! For 2D cell nodes
!              Set assembly control parameters
     IF(NDIM.EQ.2) MSYS(:,1:4)=RESHAPE((/1,2,1, 1,2,-1, 1,1,1, 2,2,1/),&
     & (/3,4/))              ! 2D
     IF(NDIM.EQ.3) MSYS=RESHAPE((/1,3,1, 1,3,-1, 1,2,1, 3,3,1, 2,3,1,  &
     & 1,1,1, 1,3,2, 2,2,1/),(/3,8/)) ! 3D
!              Set logarithmic integral coefficients(8th order)
```

```
    POSGZ=(/0.013320244161,0.079750429014,0.197871029326,        &
   & 0.354153994352,0.529458575235,0.701814529939,0.849379320441, &
   & 0.953326450056, 0.,0./)
    WEITZ=(/0.164416604728,0.237525610023,0.226841984432,         &
   & 0.175754079006,0.112924030247,0.057872210718,0.020979073742, &
   & 0.003686407104, 0.,0./)
    END SUBROUTINE BLOCK_DATA
```

5.8 Subroutine INPUT_EL

This routine reads nodal coordinates, boundary conditions, and material properties. In addition to the earlier glossary entries, brief descriptions of these variables,
in input order, are given in Appendix H.

Local variables

E: Young's modulus of elasticity.
F: Array of specified tractions.
IFLAG: Flag for prescribed tractions. If IFLAG < 0, the tractions are
              prescribed in local coordinates (ξ_1, ξ_2, n).
KNOW: Array classifying the boundary conditions into eight types.
NFIXU: Number of constrained displacement nodes.

```
    SUBROUTINE INPUT_EL(NODE,NDISP,NTRAC)
    USE VARY_ARRAYS,ONLY:CD,NUGRP,KBU,LNDB,NTGRP,KBT,LNDC,RU
    USE FIXED_VALUES; CHARACTER UCH*2,TCH*2,CDCH*12,NDCH*5
    DIMENSION KNOW(8),UCH(3),TCH(3),CDCH(3),F(NTRAC,24)
    DATA KNOW/111,0,110,1,11,100,101,10/
    DATA UCH/'UX','UY','UZ'/,TCH/'T1','T2','T3'/,CDCH/'X-COORDINATE', &
   &         'Y-COORDINATE','Z-COORDINATE'/,NDCH/' NODE'/
    IF(NDIM.EQ.2) THEN; KNOW(1)=11; KNOW(3)=10; ENDIF ! FOR 2D
    WRITE(7,'(/3X,"NODE",3A20)')(CDCH(I),I=1,NDIM)
    DO 10 N=1,NTP     ! Input nodal coordinates
    READ(5,*)M,(CD(J,M),J=1,NDIM); KBU(N)=0; NUGRP(N)=0
10  WRITE(7,'(I7,3F20.6)')M,(CD(J,M),J=1,NDIM)
!            Constrained displacement information
    READ(5,*)NFIXU     ! No. of constrained displacement nodes
    WRITE(7,'(/4X,"SPECIFIED DISP. NODES ==",I4)')NFIXU
    IF(NFIXU.NE.0)WRITE(7,'(/4X,"NODE",3X,"GROUP",4X,"MARK")')
    DO 20 IP=1,NFIXU     ! Disp. group and constrained directions
    READ(5,*)N,NUGRP(N),KBU(N)
20  WRITE(7,'(1X,2I7,I9)')N,NUGRP(N),KBU(N)
!            Multiple node information for traction-discontinuous nodes
    READ(5,*)MULTP     ! No. of traction-discontinuous nodes
    WRITE(7,'(/4X,"MULTIPLE NODES ==",I4)')MULTP
    IF(MULTP.NE.0) WRITE(7,'(/,4X,"EXTR-NODE    INITIAL-NODE")')
```

```
        DO IP=1,MULTP; READ(5,*)K,KBU(K)   ! Linked original node
        WRITE(7,'(I10,I12)')K,KBU(K); ENDDO
        NBTP=NTP-MULTP; NTF=NDIM*NBTP
        DO 40 I=1,NBP; DO 30 K=1,4*NBDM   ! Classify disp. conditions
        IF(KBU(I).NE.KNOW(K)) GOTO 30; KBU(I)=K; GOTO 40
30      CONTINUE
40      CONTINUE
        WRITE(7,'(//4X,"BOUNDARY ELEMENT AND TRACTION CONDITION :")')
        WRITE(7,'(/" ELEM GROUP MARK",8(A5,I1))')(NDCH,I,I=1,NODE)
!       Input element nodes and traction group and direction flag
        DO 60 IE=1,NBE   ! Negative IFLAG for t and n directions
        READ(5,*)M,(LNDB(M,ID),ID=1,NODE),NTGRP(M),IFLAG
        NTGRP(M)=ISIGN(NTGRP(M),IFLAG)
        DO 50 K=1,4*NBDM      ! Classify traction conditions
        IF(IABS(IFLAG).NE.KNOW(K)) GOTO 50; KBT(M)=K; GOTO 60
50      CONTINUE
60      WRITE(7,91)M,IABS(NTGRP(M)),IFLAG,(LNDB(M,ID),ID=1,NODE)
        IF(NTRAC.LE.0) GO TO 80; WRITE(7,92)(TCH(I),I=1,NDIM)
!                           Read prescribed traction values
        DO 70 N=1,NTRAC; READ(5,*)M,(F(M,J),J=1,NDIM*NODE)
        IF(NDIM.EQ.2) WRITE(7,93)M,(F(M,J),J=1,NDIM*NODE)
70      IF(NDIM.EQ.3) WRITE(7,94)M,(F(M,J),J=1,NDIM*NODE)
80      IF (NDISP.LE.0) GO TO 90; WRITE(7,95)(UCH(I),I=1,NDIM)
!                 Read prescribed displacements
        DO NE=1,NDISP; READ(5,*)M,(RU(J,M),J=1,NDIM)
        WRITE(7,'(I5,5X,3E15.6)')M,(RU(J,M),J=1,NDIM); ENDDO
90      WRITE(7,'(/5X," DIAGV        E          PR")')
        READ(5,*)DIAGV,E,PR; WRITE(7,'(1X,3E15.7/)')DIAGV,E,PR
!            Set up commonly used constants
        G=0.5*E/(1.+PR); IF(NSIG.EQ.3) PR=PR/(1.+PR)
        PR1=1.-PR; PR2=1.-2.*PR; PR3=3.-3.*PR; PR4=1.-4.*PR
        PR5=3.-4.*PR; GAM=NDIM+2.; CON=-1./(4.*NBDM*PI*PR1)
        CALL TREAT_T(NODE,NTRAC,F)   ! Treat boundary tractions
91      FORMAT(2I5,I6,I5,7I6)
92      FORMAT(//6X,'SPECIFIED TRACTION VALUES :'//1X,'GROUP',3A15)
93      FORMAT(I5,5X,1P2E16.8,2(/,10X,1P2E16.8))
94      FORMAT(I5,5X,1P3E16.8,7(/,10X,1P3E16.8))
95      FORMAT(//9X,'SPECIFIED DISPLACEMENT VALUES :'//1X,'GROUP',3A15)
        END SUBROUTINE INPUT_EL
```

5.9 Subroutine TREAT_T

This routine assigns group tractions to appropriate elements and files them on Channel 8. If the group tractions are defined with respect to the local geometry, they are first transformed into the global coordinate system. All tractions are normalized by dividing by $2G$. The global tractions t_i are obtained from the local ones t_i' from the equation

$$t_i = L_{ji}t_j' \tag{5.1}$$

where L_{ji} are the direction cosines of the local coordinates.

Local variables

CK: Array of coordinates of boundary element nodes.
COSL: Array containing L_{ij}.
F: Array of group tractions.
L: Position of the first component of nodal intrinsic coordinates.
RT: Array of element tractions.
VS: Vector of tangential direction vectors.

```
     SUBROUTINE TREAT_T(NODE,NTRAC,F)
     USE VARY_ARRAYS,ONLY:CD,NTGRP,LNDB
     USE FIXED_VALUES; REAL F(NTRAC,24),COSL(NDIM,NDIM),VS(2)
     DO 50 IE=1,NBE ! Final tractions must be in x, y, z directions
     DO 10 ID=1,8; DO 10 I=1,NDIM; IF(ID.GT.NODE) GOTO 10
     IF(NTGRP(IE).GT.0)RT(I,ID)=F(IABS(NTGRP(IE)),NDIM*(ID-1)+I)/2./G
     IF(NTGRP(IE).LT.0)CK(I,ID)=CD(I,LNDB(IE,ID))
10   IF(NTGRP(IE).LE.0.OR.ID.GT.NODE) RT(I,ID)=0.
     IF(NTGRP(IE).GE.0) GOTO 40  ! Local, if negative
     DO 30 ID=1,NODE; L=3*(ID+ID/5*4+(3-NDIM)*(ID/3)*6)-2
!         Calculate cosines of local axes with respect to global axes
     IDO=NDIM*(ID-1); CALL AXES_COS(NODE,CORDL(L),COSL,VS)
     DO 20 I=1,NDIM; DO 20 J=1,NDIM
!           Transform tractions into x, y, z directions
20   RT(I,ID)=RT(I,ID)+COSL(J,I)*F(IABS(NTGRP(IE)),IDO+J)/2./G
30   CONTINUE
40   WRITE(8,REC=IE)RT   ! Write tractions to disk
50   CONTINUE
     END SUBROUTINE TREAT_T
```

5.10 Subroutine AXES_COS

This routine calculates the direction cosines L_{ij} of a local coordinate system x_i', defined on a boundary element, with respect to the global system x_i. The relationships between these two coordinate systems are

$$x_i' = L_{ij}x_j$$
$$x_i = L_{ji}x_j' \tag{5.2}$$

The routine DSHAPE is called to calculate the normal and spatial derivatives.

Local variables

COSL: Array containing L_{ij}.
VS: Vector of tangential direction vectors.
XI: Vector of intrinsic coordinates.

```
SUBROUTINE AXES_COS(NODE,XI,COSL,VS)
USE FIXED_VALUES; REAL XI(3),COSL(NDIM,NDIM),VS(2)
CALL DSHAPE(NDIM,NBDM,NODE,XI,CK,COSB,FJCB,DN,GD,CORDL)
IF(NDIM.EQ.2) THEN
COSL(1,1)=-COSB(2); COSL(1,2)=COSB(1)   !2D tangential cosine
ELSE ! Calculate 3D tangential cosines
DO K=1,2; W=0.; DO I=1,NDIM
W=W+GD(I,K)*GD(I,K); ENDDO
VS(K)=SQRT(W)
ENDDO                    ! m(K), K=1,2
COSL(1,1:NDIM)=GD(1:NDIM,1)/VS(1)                  ! L11
COSL(2,1)=COSB(2)*COSL(1,3)-COSB(3)*COSL(1,2)      ! L12
COSL(2,2)=COSB(3)*COSL(1,1)-COSB(1)*COSL(1,3)      ! L22
COSL(2,3)=COSB(1)*COSL(1,2)-COSB(2)*COSL(1,1)      ! L23
ENDIF
!                   Assign normal Ni to L3i
COSL(NDIM,1:NDIM)=COSB(1:NDIM)
END SUBROUTINE AXES_COS
```

5.11 Subroutine SHAPEF

This routine evaluates the shape functions N_α and the parameters r^2 and r_i, where r is the distance between source and field points.

Local variables

CK: Array of element nodal coordinates.
CP: Vector of source point coordinates.
R2: r^2.
RI: Vector containing r_i.
SP: Vector of shape functions N_α.
X: Vector of the intrinsic coordinates ξ_k.

```
SUBROUTINE SHAPEF(NDIM,NDIMB,NODE,SP,X,CK,CP,RI,R2,C)
REAL SP(NODE),CK(3,NODE),CP(NDIM),RI(NDIM),C(60),X(3)
IF(NDIMB.EQ.3) GOTO 25; IF(NODE.GT.3) GOTO 10
!                   2-noded element (line element)
SP(1)=0.5*(1.-X(1)); SP(2)=0.5*(1.+X(1))
IF(NODE.EQ.2) GOTO 50
!                   3-noded element (line element)
```

```
      SP(1)=-X(1)*SP(1); SP(2)=X(1)*SP(2); SP(3)=1.-X(1)*X(1)
      GOTO 50
!                          4-noded element (quadrilateral element)
10    DO I=1,4
        SP(I)=0.25*(1.+C(3*I-2)*X(1))*(1.+C(3*I-1)*X(2))
      ENDDO
      IF(NODE.EQ.4) GOTO 50
!                 8-noded element (square element)
      DO 20 I=1,4; L=3*I-2; SP(I)=SP(I)*(C(L)*X(1)+C(L+1)*X(2)-1.)
      WL=C(L+24)*X(1)+C(L+25)*X(2)
20    SP(I+4)=0.5*(WL+1.)*(1.-(C(L+24)*X(2))**2-(C(L+25)*X(1))**2)
      GOTO 50
!                              brick elements
25    DO 30 I=1,8; K=3*I-2             ! 8-noded element
30    SP(I)=0.125*(1.+C(K)*X(1))*(1.+C(K+1)*X(2))*(1.+C(K+2)*X(3))
      IF(NODE.EQ.8) GOTO 50
      DO 35 I=1,8; K=3*I-2
35    SP(I)=SP(I)*(C(K)*X(1)+C(K+1)*X(2)+C(K+2)*X(3)-2.)
      DO 40 I=9,20; K=3*I-2            ! 20-noded element
      SP(I)=0.25*(1.+C(K)*X(1))*(1.+C(K+1)*X(2))*(1.+C(K+2)*X(3))
      SP(I)=SP(I)*(1.+(C(K)*C(K)-1.)*X(1)*X(1)+                        &
    & (C(K+1)*C(K+1)-1.)*X(2)*X(2)+(C(K+2)*C(K+2)-1.)*X(3)*X(3))
40    CONTINUE
!               Calculate r and its vector components
50    R2=0.; DO 70 I=1,NDIM; RI(I)=-CP(I)
      DO ID=1,NODE; RI(I)=RI(I)+SP(ID)*CK(I,ID); ENDDO
70    R2=R2+RI(I)*RI(I)
      END SUBROUTINE SHAPEF
```

5.12 Subroutines DSHAPE and DSHAP3D

Subroutine DSHAPE calculates the derivatives of the shape functions with respect to the intrinsic coordinates $\partial N_\alpha / \partial \xi_k$, the derivative $\partial x_i / \partial \xi_k$, and the Jacobian of the coordinate transformation. Routine DSHAP3D is not used in linear analysis.

Local variables

X: Vector containing ξ_k.

```
SUBROUTINE DSHAPE(NDIM,NDIMB,NODE,X,CK,COSB,FJCB,DN,GD,C)
REAL X(3),CK(3,NODE),DN(3,NODE),GD(3,3),COSB(NDIM),C(60),GR(3)
IF(NDIM+NDIMB.EQ.6) CALL DSHAP3D(C,X,NODE,DN) ! Plasticity only
IF(NDIM+NDIMB.EQ.6) GOTO 30; IF(NODE.GT.3) GOTO 5
DN(1,1)=-0.5; DN(1,2)=0.5           ! For 2D, 2-noded element
IF(NODE.EQ.2) GOTO 30
DN(1,1)=-0.5*(1.-2.*X(1))           ! For 2D, 3-noded element
DN(1,2)=0.5*(1.+2.*X(1)); DN(1,3)=-2.*X(1)
```

```
        GOTO 30
   5    DO 10 I=1,4; I0=3*(I-1)            ! For 3D, 4-noded element
        DN(1,I)=0.25*C(I0+1)*(1.+C(I0+2)*X(2))
  10    DN(2,I)=0.25*C(I0+2)*(1.+C(I0+1)*X(1))
        IF(NODE.EQ.4) GOTO 30
        DO 20 I=1,4; L=3*I-2; S=C(L)*X(1)+C(L+1)*X(2)-1.
        DN(1,I)=DN(1,I)*S+0.25*(1.+C(L)*X(1))*(1.+C(L+1)*X(2))*C(L)
        DN(2,I)=DN(2,I)*S+0.25*(1.+C(L)*X(1))*(1.+C(L+1)*X(2))*C(L+1)
        S=1.+C(L+24)*X(1)+C(L+25)*X(2)      ! For 3D, 8-noded element
        T=1.-(C(L+24)*X(2))**2-(C(L+25)*X(1))**2
        DN(1,I+4)=0.5*C(L+24)*T-C(L+25)*C(L+25)*X(1)*S
  20    DN(2,I+4)=0.5*C(L+25)*T-C(L+24)*C(L+24)*X(2)*S
  30    DO 50 I=1,NDIM; DO 50 J=1,NDIMB; GD(I,J)=0.; DO 50 ID=1,NODE
  50    GD(I,J)=GD(I,J)+DN(J,ID)*CK(I,ID)
        IF(NDIM.EQ.NDIMB) THEN; GOTO (51,52,53),NDIM
  51      FJCB=ABS(GD(1,1)); RETURN        ! For internal cell integral
  52      FJCB=GD(1,1)*GD(2,2)-GD(1,2)*GD(2,1); RETURN
  53      FJCB=GD(1,1)*(GD(2,2)*GD(3,3)-GD(3,2)*GD(2,3))              &
    &      -GD(1,2)*(GD(2,1)*GD(3,3)-GD(3,1)*GD(2,3)) +              &
    &       GD(1,3)*(GD(2,1)*GD(3,2)-GD(2,2)*GD(3,1)) ; RETURN ; ENDIF
        IF(NODE.GT.3) GOTO 60
        GR(1)=GD(2,1); GR(2)=-GD(1,1)                ! For 2D normals
        FJCB=SQRT(GR(1)*GR(1)+GR(2)*GR(2)); GOTO 70  ! 2D Jacobian
  60    GR(1)=GD(2,1)*GD(3,2)-GD(3,1)*GD(2,2)        ! For 3D normals
        GR(2)=GD(3,1)*GD(1,2)-GD(1,1)*GD(3,2)
        GR(3)=GD(1,1)*GD(2,2)-GD(2,1)*GD(1,2)
        FJCB=SQRT(GR(1)*GR(1)+GR(2)*GR(2)+GR(3)*GR(3))  ! 2D Jacobian
  70    COSB(1:NDIM)=GR(1:NDIM)/FJCB                 ! Normal Cosine
        END SUBROUTINE DSHAPE

        SUBROUTINE DSHAP3D(C,X,NODE,DN)
        REAL DN(3,NODE),C(60),X(3)
        DO 10 I=1,8; K=3*I-2
        DN(1,I)=0.125*C(K)*(1.+C(K+1)*X(2))*(1.+C(K+2)*X(3))
        DN(2,I)=0.125*C(K+1)*(1.+C(K)*X(1))*(1.+C(K+2)*X(3))
  10    DN(3,I)=0.125*C(K+2)*(1.+C(K)*X(1))*(1.+C(K+1)*X(2))
        IF(NODE.EQ.8) RETURN
        DO 20 I=1,8; K=3*I-2
        F=0.125*(1.+C(K)*X(1))*(1.+C(K+1)*X(2))*(1.+C(K+2)*X(3))
        S=C(K)*X(1)+C(K+1)*X(2)+C(K+2)*X(3)-2.; DO 20 J=1,3
  20    DN(J,I)=DN(J,I)*S+C(K+J-1)*F
        DO 30 I=9,20; K=3*I-2
        F=(1.+C(K)*X(1))*(1.+C(K+1)*X(2))*(1.+C(K+2)*X(3))
        S=0.25*(1.+(C(K)*C(K)-1.)*X(1)*X(1)+(C(K+1)*C(K+1)-1.)*       &
    &   X(2)*X(2)+(C(K+2)*C(K+2)-1.)*X(3)*X(3))
        DN(1,I)=C(K)*(1.+C(K+1)*X(2))*(1.+C(K+2)*X(3))*S+             &
```

```
      &  0.5*F*(C(K)*C(K)-1.)*X(1)
         DN(2,I)=C(K+1)*(1.+C(K)*X(1))*(1.+C(K+2)*X(3))*S+        &
      &  0.5*F*(C(K+1)*C(K+1)-1.)*X(2)
         DN(3,I)=C(K+2)*(1.+C(K)*X(1))*(1.+C(K+1)*X(2))*S+        &
      &  0.5*F*(C(K+2)*C(K+2)-1.)*X(3)
30    CONTINUE
      END SUBROUTINE DSHAP3D
```

5.13 Subroutine EL_COEFS

This major routine assembles the integrals of U_{ij}, T_{ij}, U_{ijk}, and T_{ijk} for boundary and interior nodes and forms the system equations $[A]\{x\} = \{y\}$. The coefficients obtained from the traction-recovery method are assembled here too.

Local variables

A:	Array containing the matrix $[A]$.
ASIG:	Array containing the matrix $[A^\sigma]$ (see Section 5.28).
IE:	Current element number.
IP:	The source (collocation) point.
NCS:	Current singular node number of current element.
NTU:	Flag for node type: 1 = boundary nodes, 2 = interior nodes, 3 = (not used – plasticity).
STRES:	Vector $\{y^\sigma\}$ in the internal stress equation (see Section 5.28).

```
      SUBROUTINE EL_COEFS(NODE)
      USE VARY_ARRAYS; USE FIXED_VALUES
      REAL CP(3),A(NDIM,NBF),ASIG(NSIG,NBF)
      WRITE(*,'(//10x," Forming ELASTIC Coefficient Matrices")')
      DO 70 IP=1,NTP    ! Loop for collocation points
            write(*,'(5x,"IP == ",i5)')ip
      NTU=1; IF(IP.GT.NBP.AND.IP.LE.NBTP) NTU=2
      IF(IP.GT.NBTP) NTU=3
      IF(NAUTO.LE.0) RLEVL=NBDM+NTU/2
      NNOD(IP)=0; IPO=NDIM*(IP-1); ISGO=NSIG*(IP-1); STRES(1:NSIG)=0.
      DO I=1,NSIG; YSIG(ISGO+I)=0.; ASIG(I,1:NBF)=0.; ENDDO
      DO 10 I=1,NDIM; CP(I)=CD(I,IP); DIAG(I,:)=0.; COEF(I,1:NBF)=0.
      DO K=1,NBF; IF(NTU.EQ.1) A(I,K)=0.; ENDDO; DIAG(I,I)=DIAGV
10    IF(IP.LE.NBTP) Y(IPO+I)=0.
      DO 50 IE=1,NBE    ! Element loop
      READ(8,REC=IE)RT; NCS=0
      DO 30 ID=1,NODE
      ND(ID)=ID; IF(NCS.NE.0) GOTO 20; NW=LNDB(IE,ID)
      IF(NW.GT.NBTP.AND.NTU.EQ.1) NW=-KBU(NW)
      IF(IABS(NW).EQ.IP) NCS=ID    ! Record singular node of element
20    HM(1:NDIM,1:NDIM,ID)=0.; GM(1:NDIM,1:NDIM,ID)=0.
      SU(1:NSIG,1:NDIM,ID)=0.; ST(1:NSIG,1:NDIM,ID)=0.
```

```
30    CK(1:NDIM,ID)=CD(1:NDIM,LNDB(IE,ID))
      IF(NCS.EQ.0.AND.NTU.EQ.3) GOTO 50
      NSP=NCS; PSYM(1:NDIM)=1.
      DO 40 ISY=1,NCSYM          ! Loop for symmetry conditions
      IF(NCS.EQ.0) THEN     ! Nonsingular element
      CALL ADAPTINT(NBDM,NODE,NSP,CP,NTU) ! Adaptive integration
      ELSE          ! Source point is one of element nodes
      IF(NTU.EQ.1) CALL SINGUHG(NODE,NCS,CP)     ! Singular element
      IF(NW.LT.0) GOTO 40; IS1=3*(NCS+NCS/5*4+(3-NDIM)*(NCS/3)*6)-2
!          Calculate boundary stresses by traction-recovery method
      CALL BDSTRS(NODE,IE,NCS,IP0,CORDL(IS1),ASIG)
      ENDIF
40    IF(ISY.LT.NCSYM) CALL SYMTRY(NDIM,NODE,NSYM(ISY+NSYM0),NCS,NSP,   &
      &ND,CSYM,CP,CK,PSYM,LSYM,1)          ! Treat symmetrical elements
!          Assembling [HM] and [GM] into system equations
      IF(NTU.EQ.1) CALL HGTOEQS(NODE,IE,IP0,A)
!               Calculate coefficients for interior points
      IF(NTU.EQ.2) CALL INNERPS(IP0,ISG0,IE,NODE,ASIG)
50    CONTINUE           ! End of element loop
      IF(NTU.EQ.2) THEN
      DO K=1,NSIG
      IF(K.LE.NDIM) WRITE(50)(COEF(K,M),M=1,NBF) ! [Au]-->(50)
      WRITE(52)(ASIG(K,M),M=1,NBF)        ! [A-sigma]-->(52)
      ENDDO
      ELSE
      DO L=1,NSIG
      YSIG(NSIG*(IP-1)+L)=STRES(INX(L))/NNOD(IP)
      WRITE(52)(ASIG(INX(L),M)/NNOD(IP),M=1,NBF)
      ENDDO
      IF(NTU.NE.1) GOTO 70   !Diagonal of [H],use rigid-body condition
      KU=KBU(IP); KPU=NSWP(KU)
      DO K=MSYS(1,KPU),MSYS(2,KPU),MSYS(3,KPU)
      A(1:NDIM,IP0+K)=DIAG(1:NDIM,K); ENDDO
      DO K=MSYS(1,KU),MSYS(2,KU),MSYS(3,KU)
      Y(IP0+1:IP0+NDIM)=Y(IP0+1:IP0+NDIM)-DIAG(1:NDIM,K)*              &
      &  RU(K,NUGRP(IP)); ENDDO
      ENDIF
      DO I=1,NDIM; WRITE(56)(A(I,J),J=1,NBF); ENDDO
70    CONTINUE   ! End of loop for collocation points
      END SUBROUTINE EL_COEFS
```

5.14 Subroutine ADAPTINT

This routine evaluates nonsingular integrals using standard Gaussian quadrature. When the source point is in close proximity to the element, an element subdivision technique (refer to Section 4.5) is used. The minimum distance (R) from the source

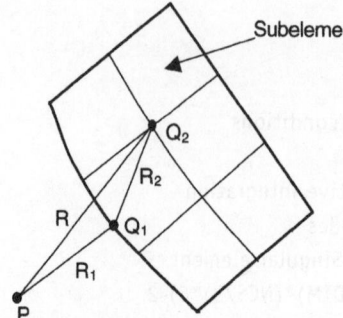

Figure 5.8: Minimum distance from source to subelements ($Q_1 \equiv \xi_{ic}$, $Q_2 \equiv \xi_i$).

point to a subelement is determined in such a way that the proximate point (*XIC*) within the original element is first found. Then, a local proximate point (*XI*) within the subelement, which gives the minimum (R_2) distance from *XIC* to this subelement (cell), is calculated. Finally, the minimum distance from the source point to the subelement can be determined from the point *XI*. Figure 5.8 depicts an example of this for an element subdivided into six subelements.

Local variables

AL:	Vector of subelement global lengths along intrinsic axes.
DGS:	Vector of subelement intrinsic lengths.
DISL:	Distance from *XIC* to a subelement.
MGAUS:	Maximum Gauss order.
NTU:	Flag for integrals: < 3, evaluate boundary integrals; > 3 evaluate domain integrals (not used – plasticity).
XIC:	Vector of the intrinsic coordinates of the proximate points.

```
      SUBROUTINE ADAPTINT(NDIMB,NODE,NSP,CP,NTU)
      USE FIXED_VALUES
      REAL CP(3),DGS(3),XI(3),XIC(3),AL(3),RI(3)
!     Determine XIC, DGS and necessary number of subelements (cells)
      MGAUS=10    ! If this number increased, change GAUSSV
      CALL CHOSEGP(NDIMB,NODE,CP,CK,DGS,XIC,AL,DISL,1,MGAUS)
!                     Non subdivision cases
      IF(NSUB(1)+NSUB(2)+NSUB(3).EQ.3) THEN
        DO IG=1,NGSS; IP=NDIMB*(IG-1)+1
          IF(NTU.LT.3) CALL INT_HG(NTU,NSP,NODE,CP,IG,GP(IP),1.)
          IF(NTU.GT.3) CALL INTSUBC(NODE,0,NTU,CP,GP(IP),IG,1.)
        ENDDO; RETURN
      ENDIF
!                     Subdivision cases
      NCOR=2**NDIMB          ! Number of element corner nodes
      DO ID=1,NCOR; CSUB(3,ID)=-1.+(CORDL(3*ID)+1.)/NSUB(3); ENDDO
      DO 70 ISUB3=1,NSUB(3)    ! No. of subcells in third direction
      DO ID=1,NCOR; CSUB(2,ID)=-1.+(CORDL(3*ID-1)+1.)/NSUB(2); ENDDO
      DO 60 ISUB2=1,NSUB(2)    ! No. of subcells in second direction
```

```
        DO ID=1,NCOR; CSUB(1,ID)=-1.+(CORDL(3*ID-2)+1.)/NSUB(1); ENDDO
        DO 50 ISUB1=1,NSUB(1)   ! No. of subcells in first direction
!           Find local proximate point XI within the subelement
        CALL MINDIST(NDIMB,NDIMB,NCOR,XIC,CSUB,1.,XI,DISL)
!           Find intrinsic coordinates RI for  XI in the original element
        CALL SHAPEF(NDIMB,NDIMB,NCOR,SHAP,XI,CSUB,CPO,RI,R2,CORDL)
!        Find distance SQRT(R2)from source point to subelement
        CALL SHAPEF(NDIM,NDIMB,NODE,SHAP,RI,CK,CP,XI,R2,CORDL)
!               Calculate Gauss orders for the subelement
        CALL CHOSEGP(NDIMB,NODE,CP,CK,DGS,XI,AL,SQRT(R2),0,MGAUS)
!                  Integrating over the subelement
        DO 40 IG=1,NGSS; IP=NDIMB*(IG-1)+1
!          Find global intrinsic coordinates XI for the Gauss point IG
        CALL SHAPEF(NDIM,NDIMB,NCOR,SHAP,GP(IP),CSUB,CPO,XI,R2,CORDL)
!          Find local Jacobian FJCBL for the subelement (cell)
        CALL DSHAPE(NDIM,NDIMB,NCOR,GP(IP),CSUB,COSB,FJCBL,DN,GD,CORDL)
!         Evaluate boundary integrals
        IF(NTU.LT.3) CALL INT_HG(NTU,NSP,NODE,CP,IG,XI,FJCBL)
!         Evaluate domain integrals for plasticity
40      IF(NTU.GT.3) CALL INTSUBC(NODE,0,NTU,CP,XI,IG,FJCBL)
!         Compute intrinsic coordinates for next subelement
50      CSUB(1,1:NCOR)=CSUB(1,1:NCOR)+DGS(1)  ! First intrinsic direction
60      CSUB(2,1:NCOR)=CSUB(2,1:NCOR)+DGS(2)  ! Second intrinsic direction
70      CSUB(3,1:NCOR)=CSUB(3,1:NCOR)+DGS(3)  ! Third intrinsic direction
        END SUBROUTINE ADAPTINT
```

5.15 Subroutine CHOSEGP

This routines calculates element dimensions and determines Gauss quadrature order. If the Gauss order exceeds a specified maximum number, then subelements are created (refer to Fig. 5.8 and Section 4.5).

Local variables

AL: Vector of subelement global lengths along intrinsic axes.

AVL: Average value of the vector AL.

CF: Array of (sub)element nodal coordinates:
global, if $NFLAG \neq 0$; intrinsic, if $NFLAG = 0$.

CP: Vector of coordinates of source point $P(x)$ if $NFLAG \neq 0$; or
coordinates of proximate intrinsic point $Q_1(\xi)$ if $NFLAG = 0$.

NFLAG: Flag: $\neq 0$, calculates "length" L_i of element, the proximate
intrinsic coordinates, minimum distance R, Gauss order m_i.
If $m_i > 10$, the number of subelements is determined.
$NFLAG = 0$: determines Gauss order m_i for subelement.

NGSP: Vector containing the Gauss integration orders
m_i ($2 \leq m_i \leq 10$).

PB: Coefficient p' (related to Gauss order).

XI: Vector containing $Q_1(\xi)$ if NFLAG $\neq$ 0; vector containing
the intrinsic coordinates of the local proximate point $Q_2(\xi)$
if NFLAG = 0.

```
      SUBROUTINE CHOSEGP(NDIMB,NODE,CP,CF,DGS,XI,AL,DISL,NFLAG,MGAUS)
      USE FIXED_VALUES
      REAL CP(3),AL(3),XI(3),CF(3,NODE),DGS(3)
      PB=SQRT(RLEVL*2./3.+0.4); IF(NFLAG.EQ.0) GOTO 30
      NSUB(1:NDIMB)=1; NGSP(1:NDIMB)=NGSIN(1:NDIMB)
      DO I=NDIMB+1,3; DGS(I)=2.; NSUB(I)=1; NGSP(I)=1; ENDDO
      IF(NAUTO.GT.0) GOTO 90  ! No subdivision
!         Calculate side lengths L_i of a element
      CALL SETGAS(NGSIN(1),1,1)  ! Set up Gauss points and weights
      AVL=0.; DO 20 ISID=1,NDIMB
      AL(ISID)=0.; DO I=1,NDIMB; IF(I.NE.ISID) XI(I)=0.0; ENDDO
      DO 10 IG=1,NGSS; XI(ISID)=GP(IG)
      CALL DSHAPE(NDIM,NDIMB,NODE,XI,CF,COSB,FJCB,DN,GD,CORDL)
!            Find Jacobian for calculation of element length
      FJCB=0.; DO I=1,NDIM; FJCB=FJCB+GD(I,ISID)*GD(I,ISID); ENDDO
10    AL(ISID)=AL(ISID)+SQRT(FJCB)*GW(IG)
20    AVL=AVL+AL(ISID)/NDIMB
!                       Calculate minimum distance R
      CALL MINDIST(NDIM,NDIMB,NODE,CP,CF,AVL,XI,DISL)
!                   Calculate Gauss integration orders
30    WFA=PB*ALOG(ABS(TOLGP)/2.)
      DO 80 I=1,NDIMB; IF(TOLGP.LT.0.) GOTO 40
      ALM=AL(I); IF(ALM.GT.3.9*DISL) ALM=3.9*DISL
      NGSP(I)=0.5*WFA/ALOG(ALM/DISL/4.)     !
      GOTO 50
40    NGSP(I)=-WFA/10.*((8./3.*AL(I)/DISL)**(3./4.)+1.)    !
50    IF(NGSP(I).LT.2) NGSP(I)=2
      IF(NFLAG.EQ.1.AND.NAUTO.LT.0) GOTO 60
      IF(NGSP(I).GT.MGAUS) NGSP(I)=MGAUS
      GOTO 80
60    IF(NGSP(I).LE.MGAUS) GOTO 70; NGSP(I)=MGAUS
!      Calculating maximum length of subelements
      IF(TOLGP.LT.0.) ALI=3./8.*DISL*(-10.*NGSP(I)/WFA-1.)**(4./3.)
!      Calculating maximum length of subelements
      IF(TOLGP.GT.0.) ALI=4.*DISL*(TOLGP/2.)**(0.5*PB/NGSP(I))
      NSUB(I)=AL(I)/ALI+0.95              ! No. of subelements
      AL(I)=AL(I)/NSUB(I)                 ! Side length of subelements
70    DGS(I)=2./NSUB(I)                   ! Step of intrinsic coordinates
80    CONTINUE
!        Set up Gauss integration coordinates and weights
90    CALL SETGAS(NGSP(1),NGSP(2),NGSP(3))
      END SUBROUTINE CHOSEGP
```

5.16 Subroutines SETGAS and GAUSSV

Subroutine SETGAS assembles Gauss points and weights into two vectors. Subroutine GAUSSV contains the numerical values.

Local variables

NGSX/Y/Z: Gauss order along element first/second/third intrinsic axes.

```
      SUBROUTINE SETGAS(NGSX,NGSY,NGSZ)
      USE FIXED_VALUES
      CALL GAUSSV(NGSX,POSGX,WEITX)
      IF(NGSY.NE.1) CALL GAUSSV(NGSY,POSGY,WEITY)
      IF(NGSZ.NE.1) CALL GAUSSV(NGSZ,POSGZ,WEITZ)
      NGSS=NGSX*NGSY*NGSZ; IPOSW=0; IPOSC=0
      DO 50 K=1,NGSZ; DO 50 J=1,NGSY; DO 50 I=1,NGSX
      IPOSW=IPOSW+1; GW(IPOSW)=WEITX(I); IPOSC=IPOSC+1
      GP(IPOSC)=POSGX(I); IF(NGSY.EQ.1) GOTO 50
      GW(IPOSW)=GW(IPOSW)*WEITY(J); IPOSC=IPOSC+1
      GP(IPOSC)=POSGY(J); IF(NGSZ.EQ.1) GOTO 50
      GW(IPOSW)=GW(IPOSW)*WEITZ(K)
      IPOSC=IPOSC+1; GP(IPOSC)=POSGZ(K)
50    CONTINUE
      END SUBROUTINE SETGAS
      SUBROUTINE GAUSSV(NGAUS,GP,GW)
      REAL GP(NGAUS),GW(NGAUS)
!     NGAUS: The number of Gauss integration points (from 2 to 10)
!     GP:    The coordinates of Gauss points
!     GW:    The weighting factors of Gauss points
      SELECT CASE(NGAUS)
      CASE(2); GP(1)=-0.57735026918962576451;  GW(1)=1.0000000000000
      CASE(3); GP(1)=-0.77459666924148337704;  GP(2)=0.0000000000000
       GW(1)=0.55555555555555555556; GW(2)=0.88888888888888888889
      CASE(4)
       GP(1)=-0.86113631159405257522; GP(2)=-0.33998104358485626480
       GW(1)=0.34785484513745385737;  GW(2)=0.65214515486254614263
      CASE(5)
       GP(1)=-0.90617984593866399280; GP(2)=-0.53846931010568309104
       GP(3)=0.00000000000000000000;  GW(1)=0.23692688505618908751
       GW(2)=0.47862867049936646804;  GW(3)=0.56888888888888888889
      CASE(6)
       GP(1)=-0.93246951420315202781; GP(2)=-0.66120938646626451366
       GP(3)=-0.23861918608319690863; GW(1)=0.17132449237917034504
       GW(2)=0.36076157304813860757;  GW(3)=0.46791393457269104739
      CASE(7)
       GP(1)=-0.94910791234275852453; GP(2)=-0.74153118559939443986
       GP(3)=-0.40584515137739716691; GP(4)=0.00000000000000000000
       GW(1)=0.12948496616886969327;  GW(2)=0.27970539148927666790
       GW(3)=0.38183005050511894495;  GW(4)=0.41795918367346938776
```

```
CASE(8)
 GP(1)=-0.96028985649753623168;  GP(2)=-0.79666647741362673959
 GP(3)=-0.52553240991632898582;  GP(4)=-0.18343464249564980494
 GW(1)=0.10122853629037625915;   GW(2)=0.22238103445337447054
 GW(3)=0.31370664587788728734;   GW(4)=0.36268378337836198297
CASE(9)
 GP(1)=-0.96816023950762608984;  GP(2)=-0.83603110732663579430
 GP(3)=-0.61337143270059039731;  GP(4)=-0.32425342340380892904
 GP(5)=0.00000000000000000000;   GW(1)=0.08127438836157441197
 GW(2)=0.18064816069485740406;   GW(3)=0.26061069640293546232
 GW(4)=0.31234707704000284007;   GW(5)=0.33023935500125976317
CASE(10)
 GP(1)=-0.97390652851717172008;  GP(2)=-0.86506336668898451073
 GP(3)=-0.67940956829902440623;  GP(4)=-0.43339539412924719080
 GP(5)=-0.14887433898163121089;  GW(1)=0.06667134430868813759
 GW(2)=0.14945134915058059315;   GW(3)=0.21908636251598204400
 GW(4)=0.26926671930999635509;   GW(5)=0.29552422471475287017
END SELECT
KGAUS=NGAUS/2; DO IGASH=1,KGAUS; JGASH=NGAUS+1-IGASH
GP(JGASH)=-GP(IGASH); GW(JGASH)=GW(IGASH); ENDDO
END SUBROUTINE GAUSSV
```

5.17 Subroutines MINDIST and IVSNR123

Subroutine MINDIST calculates the minimum distance from a point to a (sub)element, using the Newton–Raphson iterative scheme, as described in Section 4.5. Subroutine IVSNR123 calculates the inverse of a matrix of rank 3 or less.

Local variables

DISL: Minimum distance R.

GD: Array containing $[K]$.

GDT: Array containing $[K]^T[K]$.

NITER: Maximum number of iterations.

TOLE: Iteration tolerance.

XI: Intrinsic coordinates ξ_i.

XIC: Vector of the coordinates of the proximate point.

```
SUBROUTINE MINDIST(NDIME,NDIMB,NODE,CP,CF,AVL,XIC,DISL)
USE FIXED_VALUES
REAL GDT(3,3),RI(3),CP(3),CF(3,NODE),XI(3),XIC(3),REV(3,3)
TOLE=1.E-6; NITER=20; XI=0.0; DISL=AVL*1000.
DO 60 ITER=1,NITER            ! RI(i)=X(i)-CP(i)
CALL SHAPEF(NDIME,NDIMB,NODE,SHAP,XI,CF,CP,RI,RQ2,CORDL)
RM=SQRT(RQ2); IF(ABS(RM-DISL)/AVL.LT.TOLE) RETURN
!      Update minimum value of the distance R. Once R increases, return
IF(RM.LT.DISL) THEN; DISL=RM; XIC=XI; ELSE; RETURN; ENDIF
```

```
      CALL DSHAPE(NDIME,NDIMB,NODE,XI,CF,COSB,FJCB,DN,GD,CORDL)
      IF(NDIMB.EQ.NDIME) CALL IVSNR123(NDIME,GD,FJCB,GDT)
      IF(NDIMB.EQ.NDIME) GOTO 40        ! For internal cell
      IF(NDIMB.EQ.1) REV(1,1)=1./FJCB/FJCB; IF(NDIMB.EQ.1) GOTO 20
!       Multiplication of transpose of [dX(i)/dXI(K)] AND [dX(i)/dXI(K)]
      DO 10 I=1,NDIMB; DO 10 J=1,NDIMB
10    GDT(I,J)=DOT_PRODUCT(GD(1:NDIME,I),GD(1:NDIME,J))
      FJCB=GDT(1,1)*GDT(2,2)-GDT(1,2)*GDT(2,1)
      CALL IVSNR123(NDIMB,GDT,FJCB,REV)  ! Inverse of matrix
20    DO 30 I=1,NDIMB; DO 30 J=1,NDIME
30    GDT(I,J)=DOT_PRODUCT(REV(I,1:NDIMB),GD(J,1:NDIMB))
40    DO 50 I=1,NDIMB         ! Update intrinsic coordinates
50    XI(I)=XI(I)-DOT_PRODUCT(GDT(I,1:NDIME),RI(1:NDIME))  ! XI=XI+dXI
!       Scale intrinsic coordinates
      DO I=1,NDIMB; IF(ABS(XI(I)).GT.1.0) XI(I)=XI(I)/ABS(XI(I)); ENDDO
60    CONTINUE    ! End of iteration loop
      END SUBROUTINE MINDIST

      SUBROUTINE IVSNR123(NDIM,GD,FJCB,REV)
      REAL GD(3,3),REV(3,3)
!     GD:    original matrix
!     REV:   matrix inverse
!     NDIM: rank (1, 2 or 3)
!     FJCB: determinant
      GOTO(10,20,30),NDIM
10    REV(1,1)=1./FJCB
      RETURN
20    REV(1,1)=GD(2,2)/FJCB; REV(2,2)=GD(1,1)/FJCB
      REV(1,2)=-GD(1,2)/FJCB; REV(2,1)=-GD(2,1)/FJCB
      RETURN
30    REV(1,1)=(GD(2,2)*GD(3,3)-GD(3,2)*GD(2,3))/FJCB
      REV(1,2)=(GD(1,3)*GD(3,2)-GD(1,2)*GD(3,3))/FJCB
      REV(1,3)=(GD(1,2)*GD(2,3)-GD(2,2)*GD(1,3))/FJCB
      REV(2,1)=(GD(3,1)*GD(2,3)-GD(2,1)*GD(3,3))/FJCB
      REV(2,2)=(GD(1,1)*GD(3,3)-GD(1,3)*GD(3,1))/FJCB
      REV(2,3)=(GD(2,1)*GD(1,3)-GD(1,1)*GD(2,3))/FJCB
      REV(3,1)=(GD(2,1)*GD(3,2)-GD(3,1)*GD(2,2))/FJCB
      REV(3,2)=(GD(3,1)*GD(1,2)-GD(1,1)*GD(3,2))/FJCB
      REV(3,3)=(GD(1,1)*GD(2,2)-GD(1,2)*GD(2,1))/FJCB
      END SUBROUTINE IVSNR123
```

5.18 Subroutine INT_HG

This routine evaluates integrals of U_{ij} and T_{ij} over nonsingular elements, at boundary nodes. Also, it calculates U_{ij}, U_{ijk}, T_{ij}, and T_{ijk} at interior nodes to obtain interior displacements and stresses.

Local variables

AST:	Array containing the integrals of T_{ij}.
BST:	Array containing the integrals of U_{ij}.
STT:	Array containing the integrals of U_{ijk}.
SUT:	Array containing the integrals of T_{ijk}.
Z:	Vector containing $r_{,i} = r_i/r$.

```
      SUBROUTINE INT_HG(NTU,NSP,NODE,CP,IG,XI,FJCBL)
      USE FIXED_VALUES
      REAL AST(3,3),BST(3,3),SUT(6,3),STT(6,3),Z(3),CP(3),XI(3)
!              Evaluate matrices H and G
      CALL EVAL_HG(NODE,XI,GW(IG),CP,FJCBL,RN,R1,FS,Z,AST,BST,O.)
      IF(NTU.EQ.1) THEN          ! Form boundary coefficients
       CALL FORM_HG(NODE,NSP,O,AST,BST)
       RETURN
      ENDIF
!              Calculate coefficients for interior stresses
      DO 20 K=1,NDIM
      M=O; DO 10 I=1,NDIM; DO 10 J=I,NDIM; M=M+1
      TEM=NDIM*RN*(PR2*DLT(I,J)*Z(K)+PR*(DLT(I,K)*Z(J)+DLT(J,K)*Z(I))-  &
     & GAM*Z(I)*Z(J)*Z(K))+NDIM*PR*(COSB(I)*Z(J)*Z(K)+COSB(J)*Z(I)*     &
     & Z(K))+PR2*(NDIM*COSB(K)*Z(I)*Z(J)+COSB(J)*DLT(I,K)+COSB(I)*      &
     & DLT(J,K))-PR4*COSB(K)*DLT(I,J)
      SUT(INX(M),K)=-FS/R1**NDIM*TEM    !
      TEM=PR2*(Z(J)*DLT(I,K)+Z(I)*DLT(J,K)-Z(K)*DLT(I,J))+             &
     & NDIM*Z(I)*Z(J)*Z(K)
10    STT(INX(M),K)=-FS/R1**NBDM*TEM    !
      IF(NSIG.NE.4) GOTO 20
      SUT(4,K)=PR*(SUT(1,K)+SUT(2,K)); STT(4,K)=PR*(STT(1,K)+STT(2,K))
20    CONTINUE
!     Forming element coefficient matrices for internal displacements
      DO 40 ID=1,NODE; DO 40 N=1,NDIM; DO 30 M=1,NDIM
      HM(M,N,ND(ID))=HM(M,N,ND(ID))+PSYM(N)*SHAP(ID)*AST(M,N)
30    GM(M,N,ND(ID))=GM(M,N,ND(ID))+PSYM(N)*SHAP(ID)*BST(M,N)
!     Forming element coefficient matrices for internal stresses
      DO 40 M=1,NSIG
      SU(M,N,ND(ID))=SU(M,N,ND(ID))+PSYM(N)*SHAP(ID)*SUT(M,N)
40    ST(M,N,ND(ID))=ST(M,N,ND(ID))+PSYM(N)*SHAP(ID)*STT(M,N)
      END SUBROUTINE INT_HG
```

5.19 Subroutine EVAL_HG

This routine evaluates the (weighted) terms U_{ij} and T_{ij}. To normalize the system equations, the values of T_{ij} are divided by $2G$.

Local variables

FS: $\dfrac{-1}{4\pi\alpha(1-\nu)}w_1w_2|J\|J_s|$ where $\alpha = 1$ in two dimensions and $\alpha = 2$ in three dimensions.

R1:   Distance r between source and field point.

R2: r^2.

RD: r^α.

RN: $r_n = r,_i\, n_i$.

Z: Vector containing $r,_i = r_i/r$.

```
      SUBROUTINE EVAL_HG(NODE,XI,WXY,CP,SN,RN,R1,FS,Z,AST,BST,ALOGS)
      USE FIXED_VALUES
      REAL RI(3),Z(3),AST(3,3),BST(3,3),CP(3),XI(3)
!          Evaluate R**2, shape functions and Ri
      CALL SHAPEF(NDIM,NBDM,NODE,SHAP,XI,CK,CP,RI,R2,CORDL)
!          Evaluate outward normal and Jacobian
      CALL DSHAPE(NDIM,NBDM,NODE,XI,CK,COSB,FJCB,DN,GD,CORDL)
!               Evaluate Uij and Tij
      R1=SQRT(R2); RD=R1**NBDM; ALOGR=ALOGS
      IF(ALOGR.EQ.0.) ALOGR=ALOG(1./R1)      ! For 2D kernel Uij
      FS=SN*CON*WXY*FJCB; Z(1:NDIM)=RI(1:NDIM)/R1
      RN=DOT_PRODUCT(COSB(1:NDIM),Z(1:NDIM))
      DO 50 I=1,NDIM; DO 50 J=1,NDIM
!       Calculate Tij
      AST(I,J)=FS/RD*(RN*(PR2*DLT(I,J)+NDIM*Z(I)*Z(J))-                &
     & PR2*(Z(I)*COSB(J)-Z(J)*COSB(I)))
!       Calculate Uij
      IF(NDIM.EQ.2) BST(I,J)=-FS*(PR5*ALOGR*DLT(I,J)+Z(I)*Z(J))
50    IF(NDIM.EQ.3) BST(I,J)=-FS/R1*(PR5*DLT(I,J)+Z(I)*Z(J))
      END SUBROUTINE EVAL_HG
```

5.20 Subroutine FORM_HG

This subroutine assembles the integrals of U_{ij} and T_{ij} after multiplication by the shape functions. The diagonal terms of the system equations are also formed in this routine using the rigid-body translation condition. Some care is needed here to include the contributions of image elements, when allowing for symmetry.

Local variables

AST: Array containing the integrals of T_{ij}.

BST: Array containing the integrals of U_{ij}.

```
      SUBROUTINE FORM_HG(NODE,NSP,NCS,AST,BST)
      USE FIXED_VALUES; REAL AST(3,3),BST(3,3)
      DO 80 ID=1,NODE
      IF(ID.EQ.NCS) GOTO 50
```

```
!       Deal with nonsingular nodes
        DO 20 I=1,NDIM; DO 20 J=1,NDIM
        DIAG(I,J)=DIAG(I,J)-SHAP(ID)*AST(I,J)  ! Using rigid-body condition
!          Allow for symmetry:
        IF(ID.EQ.NSP) DIAG(I,J)=DIAG(I,J)+SHAP(ID)*AST(I,J)*PSYM(J)
        IF(ID.NE.NSP) HM(I,J,ND(ID))=HM(I,J,ND(ID))+                      &
     &  SHAP(ID)*AST(I,J)*PSYM(J)
20      GM(I,J,ND(ID))=GM(I,J,ND(ID))+SHAP(ID)*BST(I,J)*PSYM(J)
        GOTO 80
!       Deal with singular node
50      GM(1:NDIM,1:NDIM,ND(ID))=GM(1:NDIM,1:NDIM,ND(ID))+               &
     &  SHAP(ID)*BST(1:NDIM,1:NDIM)
80      CONTINUE
        END SUBROUTINE FORM_HG
```

5.21 Subroutine SINGUHG

This subroutine evaluates integrals of U_{ij} and T_{ij} over singular boundary elements (refer to Section 4.6). In three-dimensional problems, an element subdivision technique is used to deal with the singularities; in two-dimensional problems, logarithmic integration is employed.

Local variables

NCS: The current singular node number.
NDSID: Number of nodes of a side of a quadrilateral element.
NSB: Vector (flag): 0 = degenerate side, 1 = normal side.
NSIDE: Number of sides (4) of a quadrilateral element.
XI: Vector of the intrinsic coordinates ξ_i of a Gauss point.
XIS: Vector of intrinsic coordinates of the singular node.

```
        SUBROUTINE SINGUHG(NODE,NCS,CP)
        USE FIXED_VALUES
        DIMENSION CP(3),AST(3,3),BST(3,3),Z(3),XI(3),NSB(4),XIS(3)
!     Evaluate 2D boundary singular integrals using log integration
        IF(NDIM.EQ.2) THEN
        CALL SETGAS(NGSIN(1),1,1)  ! Gauss points for single integration
        CALL SIN2DHG(NCS,NODE,CP)
        RETURN
        ENDIF
!     Evaluate 3D boundary singular integrals using element subdivision
        CALL SETGAS(NGSIN(1),NGSIN(2),1) !Gauss pts for double integration
        NSIDE=4; NDSID=2+NODE/8  ! NDSID=2(3) linear(quadratic) elements
!          Set up flag for subelements
        DO 50 I=1,NSIDE; NSB(I)=1; DO 40 J=1,NDSID
        IF(NODEF(3*(I-1)+J).NE.NCS) GO TO 40; NSB(I)=0; GOTO 50
40      CONTINUE
```

```
50    CONTINUE
      NS0=3*(NCS+NCS/5*4-1); XIS(1:3)=CORDL(NS0+1:NS0+3)
      DO 90 ISID=1,NSIDE; IF(NSB(ISID).EQ.0) GO TO 90
!        Set up subelement for side ISID and singular point NCS
      CALL SETDSUB(NDIM,4,ISID,3,CORDL,12,NODEF,CSUB,XIS)
      DO 80 IG=1,NGSS; IP=NBDM*(IG-1)+1
!           Calculate global intrinsic coordinates XI for the Gauss point IG
      CALL SHAPEF(NDIM,NBDM,4,SHAP,GP(IP),CSUB,CP0,XI,R2,CORDL)
!           Calculate local Jacobian FJCBL for the current subelement
      CALL DSHAPE(NDIM,NBDM,4,GP(IP),CSUB,COSB,FJCBL,DN,GD,CORDL)
!           Evaluate boundary integrals
      CALL EVAL_HG(NODE,XI,GW(IG),CP,FJCBL,RN,R1,FS,Z,AST,BST,0.)
!           Form matrices [H] and [G]
80    CALL FORM_HG(NODE,0,NCS,AST,BST)
90    CONTINUE
      END SUBROUTINE SINGUHG
```

5.22 Subroutine SIN2DHG

This subroutine evaluates the singular integrals in two-dimensional problems using the method described in Section 4.6.2 for the displacement kernels.

Local variables

F: Vector containing f_1 and f_2.
NCS: The current singular node P.
XIP: ξ_p.

```
      SUBROUTINE SIN2DHG(NCS,NODE,CP)
      USE FIXED_VALUES
      REAL AST(3,3),BST(3,3),F(2),Z(3),CP(3),XI(3)
!          Evaluate nonsingular part using standard Gauss quadrature
      XIP=CORDL(3*NCS-2); L=LSYM(NCS)
      DO 30 IG=1,NGSS; DO 10 I=1,2
      IF(NCS.EQ.3) THEN      ! P is the mid-node
       F(I)=0.5*((GP(IG)-1.)*CK(I,1)+(GP(IG)+1.)*CK(I,2))-GP(IG)*CK(I,3)
      ELSE          ! P are end nodes
       IF(NODE.EQ.3) F(I)=-(2.+XIP*GP(IG))*CK(I,NCS)-              &
      & XIP*GP(IG)*CK(I,L)+2.*(1.+XIP*GP(IG))*CK(I,3) ! Quadratic element
       IF(NODE.EQ.2) F(I)=CK(I,2)-CK(I,1)     ! linear element
      ENDIF
10    CONTINUE
      ALOGR=-0.5*ALOG(F(1)*F(1)+F(2)*F(2))   !
!          Evaluate integral of nonsingular part
      CALL EVAL_HG(NODE,GP(IG),GW(IG),CP,1.,RN,R1,FS,Z,AST,BST,ALOGR)
!          Form element coefficient matrices for nonsingular part
      CALL FORM_HG(NODE,0,NCS,AST,BST)
```

```
30    CONTINUE
!         Evaluate singular part using logarithmic integration
      DO 60 IG=1,8   ! 8 Gauss points are used here
      ISUB=1  ! The subelement number
      IF(NCS.EQ.3) THEN
        XI(1)=-POSGZ(IG); FJCBL=1.
      ELSE
        XI(1)=(1.-2.*POSGZ(IG))/XIP; FJCBL=2.
      ENDIF
40    CALL SHAPEF(NDIM,NBDM,NODE,SHAP,XI,CK,CP,F,R2,CORDL)
      CALL DSHAPE(NDIM,NBDM,NODE,XI,CK,COSB,FJCB,DN,GD,CORDL)
      FS=-CON*PR5*FJCB*FJCBL*WEITZ(IG); AST(1:NDIM,1:NDIM)=0.
      BST(1:NDIM,1:NDIM)=FS*DLT(1:NDIM,1:NDIM)
!         Form element coefficient matrices for singular part
      CALL FORM_HG(NODE,0,NCS,AST,BST)
      IF(NCS.NE.3) GOTO 60
!     If P is the mid-node, calculate the second subelement
      XI(1)=POSGZ(IG); ISUB=ISUB+1; IF(ISUB.EQ.2) GOTO 40
60    CONTINUE
      END SUBROUTINE SIN2DHG
```

5.23 Subroutine SETDSUB

This routine sets up the intrinsic coordinates for degenerate subelements (refer to Section 4.6.1). Only the corner nodes of these subelements need to be established. To form a degenerate element, corner nodes on the degenerate side are assigned to the same intrinsic coordinates as the singular node. For example, the subelement (A) in Fig. 5.9 is formed by the singular node P (located at node 5) and the side with nodes 2, 6, and 3. Nodes 2 and 3 retain their original intrinsic coordinates, while nodes 1 and 4 (of that subelement) take the intrinsic coordinates of node 5.

Local variables

CSUB: Array of subelement intrinsic coordinates.

NBC: Number of edge (surface) nodes for quadrilateral (brick) cells.

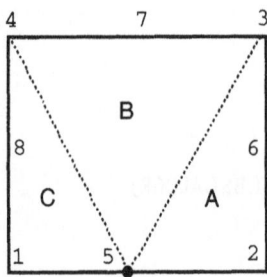

Figure 5.9: Subelements for singular point P, at node 5.

NCOR: Number of corner nodes of an element (4 or 8).

NCSD: Number of corner nodes on one side of an element (2 or 4).

```
    SUBROUTINE SETDSUB(NDIM,NCOR,ISUB,NBC,CORDL,NUM,NODEF,CSUB,XIS)
    DIMENSION CORDL(60),NODEF(NUM),CSUB(3,8),XIS(3)
    NCSD=NCOR/2
    DO 50 ID=1,NCOR          ! NCOR=4 for 2D, =8 for 3D
    DO IP=1,NCSD
    IF(NODEF(NBC*(ISUB-1)+IP).EQ.ID) GOTO 40
    ENDDO
!       Coordinates of nodes merged with the source point
    DO I=1,NDIM; CSUB(I,ID)=XIS(I); ENDDO
    GOTO 50
!       Coordinates of unmerged nodes
40  DO I=1,NDIM; CSUB(I,ID)=CORDL(3*(ID-1)+I); ENDDO
50  CONTINUE
    END SUBROUTINE SETDSUB
```

5.24 Subroutine BDSTRS

This routine evaluates and assembles the coefficients required to obtain the boundary stresses, using the traction-recovery method described in Section 4.7. After the boundary conditions are invoked, these stresses can be expressed in the form

$$\{\sigma^b\} = [A^\sigma]\{x\} + \{y^\sigma\} \tag{5.3}$$

Local variables

ASIG: Array containing matrix $[A^\sigma]$.

CL: Array containing $\partial \xi_K / \partial x'_L$.

COSL: Array of direction cosines L_{ij}.

S12:   The scalar product $(m_1 \cdot m_2)$.

STRES: Array containing vector $\{y^\sigma\}$.

TM: Vector containing the coefficients $A_{mnj\alpha}$ and B_{mnj}.

VS: Vector of the tangential direction vectors m_1 and m_2.

```
    SUBROUTINE BDSTRS(NODE,KE,NCS,IPO,XI,ASIG)
    USE VARY_ARRAYS; USE FIXED_VALUES
    REAL XI(3),ASIG(NSIG,NBF),COSL(NDIM,NDIM),CL(2,2),VS(2),TM(3)
    CALL AXES_COS(NODE,XI,COSL,VS)        ! Find L(i,j)
    CALL SHAPEF(NDIM,NBDM,NODE,SHAP,XI,CK,COSB,CL,R2,CORDL)
    IF(NDIM.EQ.2) THEN
     CL(1,1)=1./FJCB       !
    ELSE; S12=0.               !)
     DO L=1,NDIM; S12=S12+GD(L,1)*GD(L,2); ENDDO
     COSIT=S12/(VS(1)*VS(2)); SINST=SQRT(1.-COSIT*COSIT)
     CL(1,1)=1./VS(1); CL(1,2)=-COSIT/VS(1)/SINST
     CL(2,1)=0.; CL(2,2)=1./VS(2)/SINST
    ENDIF
```

```
!              Deal with displacement coefficients
     KT=KBT(KE); KPT=NSWP(KT); MP=0
     DO 70 M=1,NDIM; DO 70 N=M,NDIM; MP=MP+1
     DO 50 ID=1,NODE; NW=LNDB(KE,ND(ID))
     IF(NW.GT.NBTP) NW=KBU(NW)
     JD=NDIM*(NW-1); KU=KBU(NW); KPU=NSWP(KU); NU=NUGRP(NW); NW=MP
     CALL BSCOEF(1,NDIM,PR,PR1,PR2,DLT,COSL,CL,DN,ID,M,N,TM)
35   DO 40 J=MSYS(1,KPU),MSYS(2,KPU),MSYS(3,KPU)
     IF(ID.NE.NCS) ASIG(NW,JD+J)=ASIG(NW,JD+J)+TM(J)*PSYM(J)
40   IF(ID.EQ.NCS) ASIG(NW,JD+J)=ASIG(NW,JD+J)+TM(J)
     DO 45 J=MSYS(1,KU),MSYS(2,KU),MSYS(3,KU)
     IF(ID.NE.NCS) STRES(NW)=STRES(NW)+TM(J)*RU(J,NU)*PSYM(J)
45   IF(ID.EQ.NCS) STRES(NW)=STRES(NW)+TM(J)*RU(J,NU)
     IF(NSIG.EQ.4.AND.NW.EQ.3) THEN
      NW=4     ! FOR PLANE STRAIN
      CALL BSCOEF(2,NDIM,PR,PR1,PR2,DLT,COSL,CL,DN,ID,M,N,TM)
      GOTO 35
     ENDIF
50   CONTINUE
!                Deal with traction coefficients
     NW=MP
     CALL BSCOEF(3,NDIM,PR,PR1,PR2,DLT,COSL,CL,DN,0,M,N,TM)
55   DO 60 J=MSYS(1,KPT),MSYS(2,KPT),MSYS(3,KPT)
60   ASIG(NW,IP0+J)=ASIG(NW,IP0+J)+TM(J)
     DO 65 J=MSYS(1,KT),MSYS(2,KT),MSYS(3,KT)
65   STRES(NW)=STRES(NW)+TM(J)*RT(J,ND(NCS))
     IF(NSIG.EQ.4.AND.NW.EQ.3) THEN   ! Plane strain problems
      NW=4
      CALL BSCOEF(4,NDIM,PR,PR1,PR2,DLT,COSL,CL,DN,0,M,N,TM)
      GOTO 55
     ENDIF
70   CONTINUE
     IF(NCELL.NE.0) WRITE(55)COSL  ! for nonlinear analysis only
!            Record number of elements for average stress
     NNOD(IP0/NDIM+1)=NNOD(IP0/NDIM+1)+1
     END SUBROUTINE BDSTRS
```

5.25 Subroutine BSCOEF

This routine calculates the coefficients $A_{mnj\alpha}$ and B_{mnj} in the traction-recovery method (Section 4.7), where

$$\sigma_{mn} = A_{mnj\alpha}u_j^\alpha + B_{mnj}t_j \tag{5.4}$$

It should be noted that the coefficients $A_{mnj\alpha}$ are normalized by dividing throughout by $2G$.

Local variables

ID: Subscript α (node number).

M: Subscript M.

N: Subscript N.

NT: Flag (1–4) defining coefficients to be evaluated.

```
      SUBROUTINE BSCOEF(NT,NDIM,PR,PR1,PR2,DLT,COSL,CL,DN,ID,M,N,TM)
      REAL COSL(NDIM,NDIM),CL(2,2),DLT(3,3),DN(3,8),TM(3)
      GOTO (10,10,20,20),NT
10    DO 15 J=1,NDIM          ! To label 15 for  displacement coefficients
      IF(NT.EQ.2) TM(J)=PR/PR1*COSL(1,J)*CL(1,1)*DN(1,ID)    ! (
      IF(NT.EQ.2) GOTO 15; TM(J)=0.
      DO L=1,NDIM-1     !
      TM(J)=TM(J)+1./PR1*COSL(1,M)*COSL(1,N)*COSL(1,J)*CL(L,1)*DN(L,ID)
      IF(NDIM.EQ.2) GOTO 15
      TM(J)=TM(J)+(1./PR1*(COSL(1,M)*COSL(1,N)*PR*COSL(2,J)*CL(L,2)+      &
    & COSL(2,M)*COSL(2,N)*(COSL(2,J)*CL(L,2)+PR*COSL(1,J)*CL(L,1)))+      &
    & 0.5*(COSL(1,M)*COSL(2,N)+COSL(2,M)*COSL(1,N))*                      &
    & (COSL(1,J)*CL(L,2)+COSL(2,J)*CL(L,1)))*DN(L,ID)    !
      ENDDO
15    CONTINUE
      RETURN
20    DO 30 J=1,NDIM         ! To label 30 for traction coefficients
      IF(NT.EQ.4) THEN; TM(J)=PR/PR1*COSL(2,J)    !
      GOTO 30; ENDIF
      TM(J)=(COSL(NDIM,M)*COSL(1,N)+COSL(1,M)*COSL(NDIM,N))*COSL(1,J)
      IF(NDIM.EQ.2) THEN   !
       TM(J)=TM(J)+(DLT(M,N)-PR2/PR1*COSL(1,M)*COSL(1,N))*COSL(2,J)
       GOTO 30
      ENDIF    !
      TM(J)=TM(J)+(COSL(2,M)*COSL(3,N)+COSL(3,M)*COSL(2,N))*COSL(2,J)+    &
    & (PR/PR1*DLT(M,N)+PR2/PR1*COSL(3,M)*COSL(3,N))*COSL(3,J)
30    CONTINUE
      END SUBROUTINE BSCOEF
```

5.26 Subroutine SYMTRY

This routine generates symmetrical element data, including nodal coordinates, node numbers, and symmetry multipliers (see Section 4.8).

Local variables

CK: Array of coordinates (x_i^s) of the image nodes, where

$$x_i^s = 2x_i^c - x_i^o \quad \text{when } i = IC,$$
$$x_i^s = x_i^o \qquad \text{when } i \neq IC,$$

and x_i^c, x_i^o defines the symmetry planes and original nodes.

CQ: Array temporarily containing nodal coordinates.
DST12:   Distance between nodes 1 and 2.
IC: Symmetry plane identifier.
NCS: Singular nodal number of current symmetric element. If
                 NSP = 0, then NCS = 0. NSP ≠ 0, then NCS may be zero
                 (if source P not located on symmetry plane) or nonzero (if P is
                 on symmetry plane).
NP: Flag for determination of the symmetry multiplier PSYM: = 1
                 for displacement and traction, = 2 for stresses.
NSP: Node number corresponding to the original singular node. If
                 zero, the original element does not include the source point.
PSYM: Vector of symmetry multipliers. Set to −1 in the symmetric
                 direction and to +1 in other directions (cf. Section 4.8).

```
     SUBROUTINE SYMTRY(NDIM,NODE,IC,NCS,NSP,ND,CSYM,CP,CK,PSYM,NSWP,NP)
     DIMENSION ND(NODE),CSYM(3),CP(3),CK(3,NODE),PSYM(6),NSWP(NODE),   &
     & CQ(3,20)
     DST12=0.; DO I=1,NDIM; DST12=DST12+(CK(I,1)-CK(I,2))**2; ENDDO
!              Determine coordinates of the symmetric element nodes
     CQ(:,1:NODE)=CK(:,1:NODE); CQ(IC,:)=2.*CSYM(IC)-CQ(IC,:); NCS=0
!         Determine symmetric element nodes and singular node
     DO 50 ID=1,NODE; NID=NSWP(ID)
     IF(NODE.EQ.3.AND.ID.EQ.3) NID=3 !No swap for 3rd node of element
     CK(1:NDIM,ID)=CQ(1:NDIM,NID); DSTP=0.
     DO I=1,NDIM; DSTP=DSTP+(CK(I,ID)-CP(I))*(CK(I,ID)-CP(I)); ENDDO
     IF(SQRT(DSTP/DST12).LT.0.01) NCS=ID  ! Determine singular node
     ND(ID) = NSWP(ND(ID)) ! symmetric element nodes
50   IF(NODE.EQ.3.AND.ID.EQ.3) ND(ID)=3
!              Determine symmetry multipliers for displacement and traction
     IF(NP.EQ.1) PSYM(IC) = -PSYM(IC)
     IF(NODE.EQ.3.AND.NSP.EQ.3) RETURN
!              Determine the corresponding node of original singular node
     IF(NSP.NE.0) NSP=NSWP(NSP); IF(NP.EQ.1) RETURN
!              Determine symmetry multipliers for stresses
     DO 60 I=1,NDIM-1; DO 60 J=I+1,NDIM
     ISC=I+J-1+IDIM(I+J,4)
60   IF(I.EQ.IC.OR.J.EQ.IC) PSYM(ISC)=-PSYM(ISC)
     END SUBROUTINE SYMTRY
```

5.27 Subroutine HGTOEQS

This routine assembles the system equations $[A]\{x\} = \{y\}$. The global arrays
MSYS and NSWP, initialized in BLOCK_DATA, control the assembly of the
system equations.

Local variables

A: Array containing system matrix $[A]$.
KT: The type of traction boundary condition.
KU: The type of displacement boundary condition.
KCT: The counterpart of KT.
KCU: The counterpart of KU.
NU: Group number of the specified displacements.
Y: Vector containing known vector $\{y\}$.

```
SUBROUTINE HGTOEQS(NODE,IE,IO,A)
USE VARY_ARRAYS,ONLY:Y,NUGRP,RU,KBT,KBU,LNDB; USE FIXED_VALUES
REAL A(NDIM,NBF)
DO 50 ID=1,NODE; M=LNDB(IE,ID)
IF(M.GT.NBP) M=KBU(M) ! extra nodes at traction discontinuities
J0=NDIM*(M-1); KU=KBU(M); KT=KBT(IE); NU=NUGRP(M)
KCU=NSWP(KU); KCT=NSWP(KT)
DO J=MSYS(1,KU),MSYS(2,KU),MSYS(3,KU)
  Y(IO+1:IO+NDIM)=Y(IO+1:IO+NDIM)-HM(1:NDIM,J,ID)*RU(J,NU)
ENDDO
DO J=MSYS(1,KCU),MSYS(2,KCU),MSYS(3,KCU)
  A(1:NDIM,J0+J)=A(1:NDIM,J0+J)+HM(1:NDIM,J,ID)
ENDDO
DO J=MSYS(1,KT),MSYS(2,KT),MSYS(3,KT)
  Y(IO+1:IO+NDIM)=Y(IO+1:IO+NDIM)+GM(1:NDIM,J,ID)*RT(J,ID)
ENDDO
DO J=MSYS(1,KCT),MSYS(2,KCT),MSYS(3,KCT)
  A(1:NDIM,J0+J)=A(1:NDIM,J0+J)-GM(1:NDIM,J,ID)
ENDDO
50  CONTINUE
END SUBROUTINE HGTOEQS
```

5.28 Subroutine INNERPS

This routine forms the matrices $[A^u]$, $[A^\sigma]$ and the vectors $\{yu\}$, $\{y^\sigma\}$ for interior points, as defined in the following equations. These arise from the discretized form of the boundary integral equations for displacements (Eq. 3.22) and stresses (Eq. 3.33) in the interior of the domain. Thus, the interior displacements are obtained from the equation

$$\{u^I\} = [G^I]\{t\} - [H^I]\{u\} \qquad (5.5)$$

where the superscript I denotes interior quantities. After substitution of the boundary conditions, we obtain

$$\{u^I\} = [A^u]\{x\} + \{y^u\} \qquad (5.6)$$

Similarly, for interior stresses we obtain

$$\{\sigma^I\} = [A^\sigma]\{x\} + \{y^\sigma\} \qquad (5.7)$$

Boundary conditions are treated analogously to HGTOEQS.

Local variables

ASIG:	Array containing $[A^\sigma]$.
COEF:	Array containing $[A^u]$.
Y:	Vector containing $\{y^u\}$ for interior displacements.
YSIG:	Vector containing $\{y^\sigma\}$.

```
SUBROUTINE INNERPS(IPO,ISGO,IE,NODE,ASIG)
USE VARY_ARRAYS; USE FIXED_VALUES
REAL ASIG(NSIG,NBF)
KT=KBT(IE); KCT=NSWP(KT)   ! For traction conditions
DO 50 ID=1,NODE; M=LNDB(IE,ID)
IF(M.GT.NBTP) M=KBU(M) ! extra nodes at traction discontinuities
JD=NDIM*(M-1); KU=KBU(M); KCU=NSWP(KU); NU=NUGRP(M)
DO 50 MP=1,NSIG
DO K=MSYS(1,KU),MSYS(2,KU),MSYS(3,KU)
  IF(MP.LE.NDIM) Y(IPO+MP)=Y(IPO+MP)-HM(MP,K,ID)*RU(K,NU)
  YSIG(ISGO+MP)=YSIG(ISGO+MP)-SU(MP,K,ID)*RU(K,NU)
ENDDO
DO K=MSYS(1,KCU),MSYS(2,KCU),MSYS(3,KCU)
  IF(MP.LE.NDIM) COEF(MP,JD+K)=COEF(MP,JD+K)-HM(MP,K,ID)
  ASIG(MP,JD+K)=ASIG(MP,JD+K)-SU(MP,K,ID)
ENDDO
DO K=MSYS(1,KT),MSYS(2,KT),MSYS(3,KT)
  IF(MP.LE.NDIM) Y(IPO+MP)=Y(IPO+MP)+GM(MP,K,ID)*RT(K,ID)
  YSIG(ISGO+MP)=YSIG(ISGO+MP)+ST(MP,K,ID)*RT(K,ID)
ENDDO
DO K=MSYS(1,KCT),MSYS(2,KCT),MSYS(3,KCT)
  IF(MP.LE.NDIM) COEF(MP,JD+K)=COEF(MP,JD+K)+GM(MP,K,ID)
  ASIG(MP,JD+K)=ASIG(MP,JD+K)+ST(MP,K,ID)
ENDDO
50  CONTINUE
END SUBROUTINE INNERPS
```

5.29 Subroutine EL_SOLVE

This routine calls the solver of the system equations $[A]\{x\} = \{y\}$ and then calculates the stresses at boundary and interior nodes.

```
SUBROUTINE EL_SOLVE
USE VARY_ARRAYS; USE FIXED_VALUES
REAL A(NBF,NBF+1); DOUBLE PRECISION SS
WRITE(*,'(//11x,"Solving Elastic System Equations")')
```

```
      REWIND(56)    ! Containing [A]
      DO I=1,NBF; READ(56)(A(I,J),J=1,NBF); ENDDO   ! Take back [A]
      A(1:NBF,NBF+1)=Y(1:NBF)     ! Assign y to the last column of [A]
!            Find inverse matrix of [A] and boundary unknowns x
      CALL INVSOLVR(NBF,NBF+1,A,NBF,0)
!                 Store [A]⁻¹ in channel (56) for use in plastic part
      REWIND(56); DO I=1,NBF; WRITE(56)(A(I,J),J=1,NBF); ENDDO
!            Transfer boundary unknowns to vector {x}
      X(1:NBF)=A(1:NBF,NBF+1); REWIND (50)
!         Calculate interior displacements
      DO 40 IP=NBP+1,NBTP; IP0=NDIM*(IP-1); DO 40 J=1,NDIM
      READ(50)(COEF(1,K),K=1,NBF)        ! Recall [Au]
      SS=0.0D0; DO K=1,NBF; SS=SS+DBLE(COEF(1,K))*DBLE(X(K)); ENDDO
40    X(IP0+J)=Y(IP0+J)+SS
!         Calculate stresses and store them in array ESTRS
      REWIND(52)
      DO 60 IP=1,NTP; DO 60 I=1,NSIG
      READ(52)(COEF(1,K),K=1,NBF)  ! Recall [A-sigma]
      SS=0.; DO L=1,NBF; SS=SS+DBLE(COEF(1,L))*DBLE(X(L)); ENDDO
60    ESTRS(I,IP)=YSIG(NSIG*(IP-1)+I)+SS
      END SUBROUTINE EL_SOLVE
```

5.30 Subroutine INVSOLVR

This subroutine evaluates the inverse of a matrix and/or solves the associated set of equations $[A]\{x\} = \{y\}$ using the Gauss–Jordan method with partial pivoting. The vector $\{y\}$ is the $(n + 1)$-th column of $[A]$. After the computation, $[A]$ contains the inverse matrix and the vector $\{x\}$.

Local variables

A: Array containing the matrix $[A]$.
INDIC: Flag $(-1 \to [A]^{-1}, 0 \to [A]^{-1}$ & $\{x\}; +1 \to \{x\})$
N: Number of equations.
NCOL: Number of columns in $[A]$.
NROW: Number of rows in $[A]$.

```
      SUBROUTINE INVSOLVR(NROW,NCOL,A,N,INDIC)
      DIMENSION A(NROW,NCOL),W(N),IROW(N)
      EPS=1.E-8  ! The tolerance of the minimum pivot
      DO 40 K=1,N
      KM1=K-1; PIVOT=0.
      DO 20 I=1,N; IF(K.EQ.1) GOTO 10
      DO ISCAN=1,KM1; IF(I.EQ.IROW(ISCAN)) GOTO 20; ENDDO
10    IF(ABS(A(I,K)).LE.ABS(PIVOT)) GOTO 20
      PIVOT=A(I,K); IROW(K)=I
20    CONTINUE
      IF(ABS(PIVOT).GT.EPS) GOTO 30; STOP 9999
```

```
30   IROWK=IROW(K)
     A(IROWK,1:NCOL)=A(IROWK,1:NCOL)/PIVOT; A(IROWK,K)=1./PIVOT
     DO 40 I=1,N; AIK=A(I,K); IF(I.EQ.IROWK) GOTO 40
     A(I,1:NCOL)=A(I,1:NCOL)-AIK*A(IROWK,1:NCOL); A(I,K)=-AIK/PIVOT
40   CONTINUE
     IF(INDIC.LT.0) GOTO 60
     DO 50 IX=N+1,NCOL; W(1:N)=A(IROW(1:N),IX); DO 50 I=1,N
50   A(I,IX)=W(I)
     IF(INDIC.GT.0) RETURN
60   DO 70 J=1,N; W(1:N)=A(IROW(1:N),J); DO 70 I=1,N
70   A(I,J)=W(I)
     DO 80 I=1,N; W(IROW(1:N))=A(I,1:N); DO 80 J=1,N
80   A(I,J)=W(J)
     END SUBROUTINE INVSOLVR
```

5.31 Subroutines OUTPUT and SIGTITL

OUTPUT writes displacements, tractions, and stresses; SIGTITL writes titles.
In OUTPUT, two functions are defined: LOUT and INC. The first extracts the
numbers I, J, K, and L from KOUT, while the second determines whether the
numbers are integer multiples of the variable INCRS.

Local variables

EQBCK: Equivalent back stress (not used – plasticity).

EQSTR: Equivalent stresses (not used – plasticity).

INCRS: Output flag. In linear analyses, it is set to *unity*.

KOUT: Output flag:

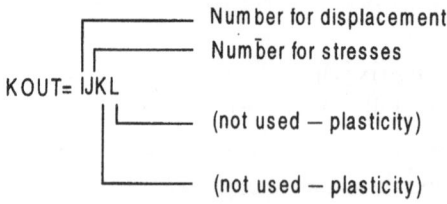

        In general, the numbers I, J, K and L can take values from 1 to 9.
        In linear analyses, $I = J = 1$ and $K = L = 0$.

STRES: Vector of stresses.

UTV: Vector of boundary displacements and tractions.

```
SUBROUTINE OUTPUT(NODE,KOUT,INCRS,FSCAL)
USE VARY_ARRAYS; USE FIXED_VALUES
CHARACTER PCH*5,UCH*2,TCH*2; DIMENSION UCH(3),TCH(3),UTV(6)
DATA PCH/' NODE'/,UCH/'UX','UY','UZ'/,TCH/'TX','TY','TZ'/
LOUT(N)=KOUT/N-(KOUT/(N*10))*10
INC(N)=((INCRS/N)*N)/INCRS+1/INCRS
```

```
      M=LOUT(1000); IF(M.NE.0) M=INC(M);IF(M.EQ.0) GOTO 40
!                 Print displacements and tractions
      WRITE(7,'(/A5,3A10,3A13)')PCH,(UCH(I),I=1,NDIM),(TCH(I),I=1,NDIM)
      DO 30 IP=1,NBTP; IP0=NDIM*(IP-1)
      IF(IP.GT.NBP) THEN         ! For interior points
       WRITE(7,92)IP,(X(IP0+I),I=1,NDIM); GOTO 30
      ENDIF
      DO 10 IE=1,NBE; DO 10 ID=1,NODE; IF(LNDB(IE,ID).NE.IP) GOTO 10
      LE=IE; LD=ID; IF(NTGRP(LE).GT.0) GOTO 20
10    CONTINUE
20    READ(8,REC=LE)RT    ! Recall specified tractions
      DO J=1,NDIM; UTV(J)=X(IP0+J); UTV(3+J)=RT(J,LD)*2.*G*FSCAL; ENDDO
      KU=KBU(IP)  ! Retrieving the specified displacements
      DO J=MSYS(1,KU),MSYS(2,KU),MSYS(3,KU)
       UTV(J)=RU(J,NUGRP(IP))*FSCAL; UTV(3+J)=X(IP0+J)*2.*G
      ENDDO
      WRITE(7,92)IP,(UTV(J),J=1,NDIM),(UTV(3+J),J=1,NDIM)
30    CONTINUE
!                       Print stresses
40    M=LOUT(100); IF(M.NE.0) M=INC(M); IF(M.EQ.0) GOTO 60
      WRITE(7,'(//25X,"STRESSES:")'); CALL SIGTITL
      DO 50 IP=1,NTP; STRES(1:NSIG)=(ESTRS(:,IP)+BACKS(:,IP))*2.*G
50    WRITE(7,91)IP,(STRES(I),I=1,NSIG)
!                       Print back stresses
60    M=LOUT(10); IF(M.NE.0) M=INC(M); IF(M.EQ.0) GOTO 70
      WRITE(7,'(//22X,"BACK-STRESSES:")'); CALL SIGTITL
      DO IP=1,NTP; WRITE(7,91)IP,(BACKS(I,IP),I=1,NSIG); ENDDO
!        Print equivalent plastic strain, stress and back stress
70    M=LOUT(1); IF(M.NE.0) M=INC(M); IF(M.EQ.0) RETURN; WRITE(7,93)
      DO 80 IP=1,NTP; STRES(1:NSIG)=ESTRS(:,IP)+BACKS(:,IP)
      CALL PL_FLOW(STRES,EQSTR,DF,0)         ! Find equiv. stress
      CALL PL_FLOW(BACKS(1,IP),EQBCK,DF,0)  ! Find equiv. back stress
80    WRITE(7,94)IP,EPSTN(IP),EQSTR*2.*G,EQBCK*2.*G
91    FORMAT(I4,1X,1P3E14.6,1P4E12.4)
92    FORMAT(I4,1X,3F11.6,1P3E14.6)
93    FORMAT(//,' NODE   EQ-P-STRAIN    EQ-STRESS   EQ-BACK-STRESS')
94    FORMAT(I5,1X,1P3E15.6)
      END SUBROUTINE OUTPUT

      SUBROUTINE SIGTITL
      USE VARY_ARRAYS; USE FIXED_VALUES
      CHARACTER PCH*5,SS*2; DIMENSION SS(6)
      DATA PCH/' NODE'/,SS/'XX','YY','ZZ','XY','YZ','ZX'/
      IF(NSIG.EQ.6) WRITE(7,'(/A5,6(4X,"STRESS-",A2))')        &
    & PCH,(SS(I),I=1,NSIG)
      IF(NSIG.EQ.4) WRITE(7,'(/A5,1X,4(4X,"STRESS-",A2))')        &
```

```
& PCH,(SS(I),I=1,2),SS(4),SS(3)
  IF(NSIG.EQ.3) WRITE(7,'(/A5,1X,3(4X,"STRESS-",A2))')      &
& PCH,(SS(I),I=1,2),SS(4)
  END SUBROUTINE SIGTITL
```

5.32 Closure

From the standpoint of the programmer, this chapter constitutes a (necessarily) abbreviated manual, although some effort has been expended to ensure that its structure facilitates the programmer's principal task of deciphering the code. In addition, the explanatory matter is intended to develop an understanding of how the theoretical and numerical formulations of the previous chapters can be translated into concrete form. Before running the program in earnest, a review of the description of the input data in Appendix H, and of some example problems described in the following chapter, will prove worthwhile.

CHAPTER SIX

Linear Applications

6.1 Introduction

In this chapter, we present some applications of the computer code BEMECH described in Chapter Five. As well as benchmark tests, we also demonstrate some reasonably realistic three-dimensional applications. These examples also provide an opportunity to illustrate the format of the input data, some of which are reproduced here. The input and output data for most of these problems are contained in the electronic files supplied with this book: these should facilitate program familiarization and validation.

6.2 Thick-Walled Cylinder under Internal Pressure

A thick cylinder, with an internal diameter of 200 units and an external diameter of 400 units, is subjected to an internal pressure of ten units under plane strain conditions as depicted in Fig. 6.1a. The elastic material properties are: Young's modulus of elasticity $E = 21000$ and Poisson's ratio $\nu = 0.3$.

Exploiting symmetry, only a quarter of the cylinder needs to be discretized into boundary elements. The discretization scheme consists of just twelve quadratic boundary elements, defined by twenty-six nodes, as depicted in Fig. 6.1b. No elements need to be defined on the axes of symmetry. However, a further three nodes are defined in the interior, in order to trace the internal stress and displacement distributions. The input data (also contained in the file CYLE2D.DAT) for this problem are as follows:

```
Thick-cylinder: plane strain elastic analysis
4     3      26    29    12   0   1   0    4   -6
 -0.00001        0.0      0.0
  1        0.00     100.00
  2       13.40      99.02
  3       26.15      96.48
  4       38.26      92.38
```

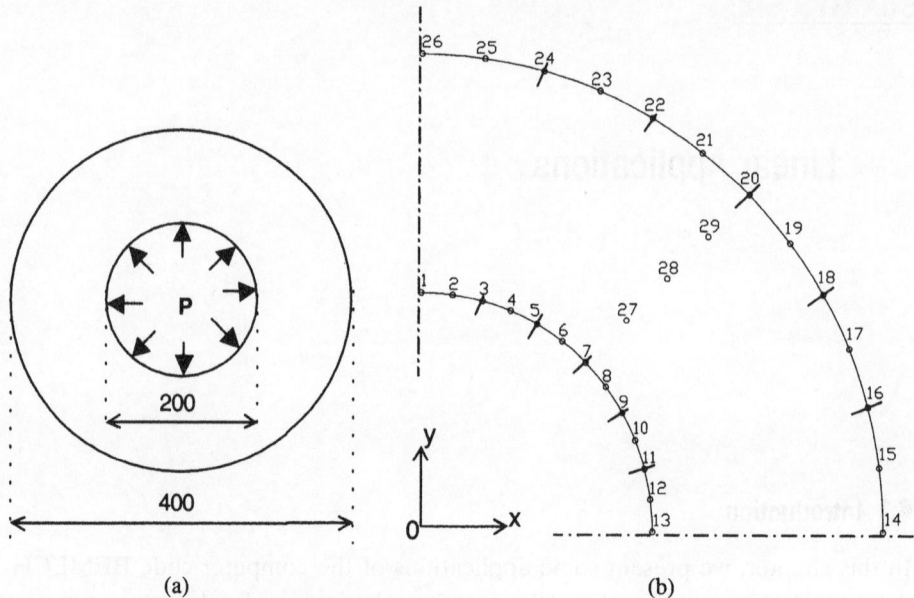

Figure 6.1: Thick cylinder under internal pressure: (a) schematic, (b) discretization.

5	49.72	86.72
6	60.54	79.49
7	70.71	70.71
8	79.49	60.54
9	86.72	49.72
10	92.38	38.26
11	96.48	26.15
12	99.02	13.40
13	100.00	0.00
14	200.00	0.00
15	198.05	26.80
16	192.97	52.31
17	184.77	76.53
18	173.44	99.45
19	158.99	121.08
20	141.42	141.42
21	121.08	158.99
22	99.45	173.44
23	76.53	184.77
24	52.31	192.97
25	26.80	198.05
26	.00	200.00
27	88.38	88.38
28	106.06	106.06
29	123.74	123.74
0		
0		

1	1	3	2	1	-11
2	3	5	4	1	-11
3	5	7	6	1	-11
4	7	9	8	1	-11
5	9	11	10	1	-11
6	11	13	12	1	-11
7	14	16	15	0	11
8	16	18	17	0	11
9	18	20	19	0	11
10	20	22	21	0	11
11	22	24	23	0	11
12	24	26	25	0	11

```
1      0.00 -10.00  0.00 -10.00   0.00 -10.00
0.00   21000.00 0.3
```

Edited extracts of the output data are as follows (the complete output data may be inspected in the file CYLE2D.OUT):

```
Thick-cylinder: plane strain elastic analysis
NSIG NODE NBP NTP NBE NCELL NTRAC NDISP KSYM  NAUTO
   4    3   26  29   12     0     1     0    4     -6
```

Gauss tolerance	==	-0.00001	
Symmetric coordinates	==	0.00	0.00
Specified disp. nodes	==	0	
Multiple nodes	==	0	

TRACTIONS:

Group	T1	T2
1	0.00	-10.00
	0.00	-10.00
	0.00	-10.00

Diagonal	E-modulus	P-ratio
0.00	21000.00	0.3

Displacements & Tractions:

Node	Ux	Uy	Tx	Ty
1	0.0000	0.0907	0.140	9.999
2	0.0121	0.0899	1.330	9.911
3	0.0236	0.0876	2.579	9.661
4	0.0347	0.0838	3.826	9.238
5	0.0452	0.0786	5.008	8.655
6	0.0550	0.0722	6.067	7.949
7	0.0641	0.0641	7.169	6.970
8	0.0722	0.0550	7.949	6.067
9	0.0786	0.0452	8.655	5.008
10	0.0838	0.0347	9.238	3.826
11	0.0876	0.0236	9.661	2.579
12	0.0899	0.0121	9.911	1.330
13	0.0907	0.0000	9.999	0.140

14	0.0577	0.0	0.0	0.0
15	0.0572	0.0077	0.0	0.0
16	0.0557	0.0151	0.0	0.0
17	0.0533	0.0220	0.0	0.0
18	0.0501	0.0286	0.0	0.0
19	0.0459	0.0349	0.0	0.0
20	0.0408	0.0408	0.0	0.0
21	0.0349	0.0459	0.0	0.0
22	0.0286	0.0501	0.0	0.0
23	0.0220	0.0533	0.0	0.0
24	0.0151	0.0557	0.0	0.0
25	0.0077	0.0572	0.0	0.0
26	0.0000	0.0577	0.0	0.0
27	0.0539	0.0539		
28	0.0476	0.0476		
29	0.0435	0.0435		

Stresses:

Node	S-xx	S-yy	S-xy	S-zz
1	16.701	-9.996	-0.140	2.011
2	16.110	-9.529	-3.505	1.974
3	14.893	-8.226	-6.645	2.000
4	12.885	-6.073	-9.479	2.043
5	9.977	-3.311	-11.559	1.999
6	6.796	-0.214	-12.820	1.974
7	3.351	3.351	-13.345	2.010
8	-0.214	6.796	-12.820	1.974
9	-3.311	9.977	-11.559	1.999
10	-6.073	12.885	-9.479	2.043
11	-8.226	14.892	-6.645	2.000
12	-9.529	16.110	-3.505	1.974
13	-9.996	16.701	-0.140	2.011
14	.001	6.675	-0.0	2.003
15	.118	6.565	-0.881	2.005
16	.443	6.221	-1.660	1.999
17	.970	5.654	-2.342	1.987
18	1.671	4.993	-2.889	1.999
19	2.460	4.223	-3.223	2.005
20	3.338	3.338	-3.337	2.003
21	4.223	2.460	-3.223	2.005
22	4.993	1.671	-2.889	1.999
23	5.654	.970	-2.342	1.987
24	6.221	.443	-1.660	1.999
25	6.565	.118	-0.881	2.005
26	6.675	.001	0.0	2.003

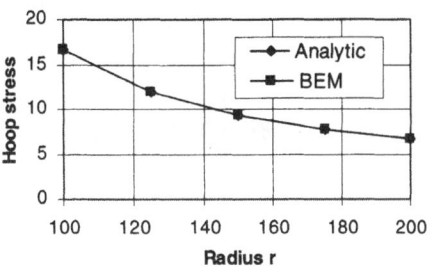

Figure 6.2: Hoop stress and radial displacement in thick cylinder.

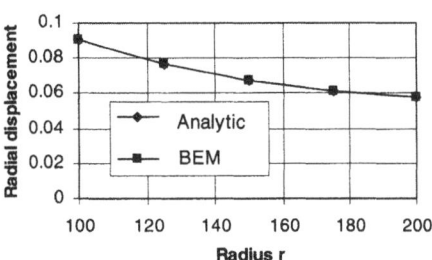

27	3.328	3.328	-8.525	1.997
28	3.333	3.333	-5.919	2.000
29	3.334	3.334	-4.353	2.000

Figure 6.2 shows plots of the radial displacement and hoop (circumferential) stress distributions along the radial direction. The excellent agreement between the numerical and analytical solutions (Lubliner, 1990) is evident.

6.3 Circular Rigid Foundation on a Semi-Infinite Medium

The smooth circular rigid foundation (with radius $R = 1$), depicted in Fig. 6.3a, rests on an elastic half-space (defined by its Young's modulus of elasticity E, assumed to be unity, and Poisson's ratio ν) and is subjected to unit vertical displacement. Using quadrantal symmetry, the foundation itself is discretized using 45 quadratic elements, as shown in Fig. 6.3b. Small elements are used near the edge of the foundation because of the expected traction singularity there. Outside the foundation, the infinite (free) surface is truncated at a radius of 50 units and discretized using just 5 quadratic boundary elements in the radial direction. These increase in dimension away from the foundation. Thus, in total, the discretization consists of 75 elements, defined by 260 nodes. The input and output data for this example may be found in the files PLATE3D.DAT and PLATE3D.OUT, respectively.

Table 6.1 records the computed values of the reaction force F, for three values of Poisson's ratio. Despite the difficulties of resolving the edge singularity, the error is in no case greater than 0.7%, in comparison with the reported analytical solution (Poulos & Davis, 1974).

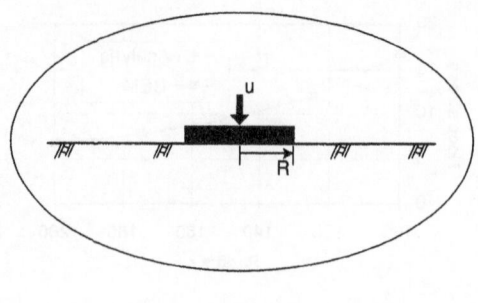

(a)

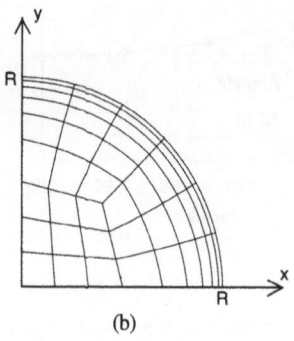

(b)

Figure 6.3: Circular rigid foundation: (a) schematic, (b) discretization.

Further examination of the edge singularity, as exemplified by the plot (Fig. 6.4) of the distribution of the vertical traction t_z along the radial direction (r) for the case $\nu = 0.3$, shows that the tractions rise monotonically, as expected, toward the edge, using the stated discretization scheme. However, numerical trials show that if one uses a coarse mesh, the singularity can contaminate the (traction) solutions toward the interior of the foundation. Nevertheless, it appears that the vertical reaction force can be determined satisfactorily even with a coarse mesh.

6.4 A Three-Dimensional Machine Component

A practical application of the boundary element method for the analysis of a machine component (an axle bearing) is shown in Fig. 6.5. The thickness of the component (in the z direction) is 20 mm. The normal pressure on the bearing is distributed over the inner upper half of the bearing surface according to the equation $T_n = Fz\cos\theta$ (GPa), where $F = 0.061$, $-10 \leq z \leq 10$, and $-90° \leq \theta \leq 90°$.

Table 6.1. Reaction force induced by unit displacement			
ν	Current	Analytical	Error (%)
0.0	2.015	2.	0.7
0.3	2.207	2.198	0.45
0.5	2.671	2.663	0.3

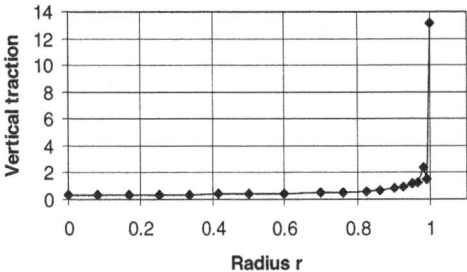

Figure 6.4: Vertical traction distribution along radius.

That is, the pressure varies sinusoidally in the hoop direction (maximum vertically upward) and linearly in the z-direction, resulting in out-of-plane bending. Full fixity is provided along the boundary (identified by the thicker line weight) at the lower right, that is, over the half-cylinder cutout and the adjacent plane surface

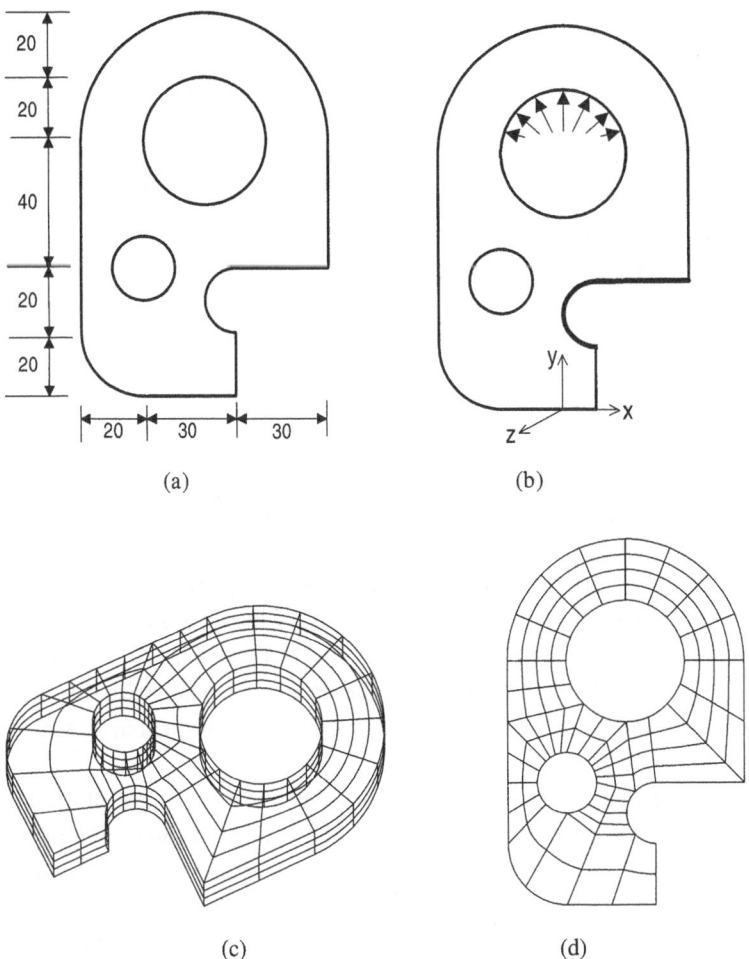

Figure 6.5: Machine component: (a) dimensions (in mm), (b) boundary conditions, (c) 3D view of mesh, and, (d) plane view of mesh.

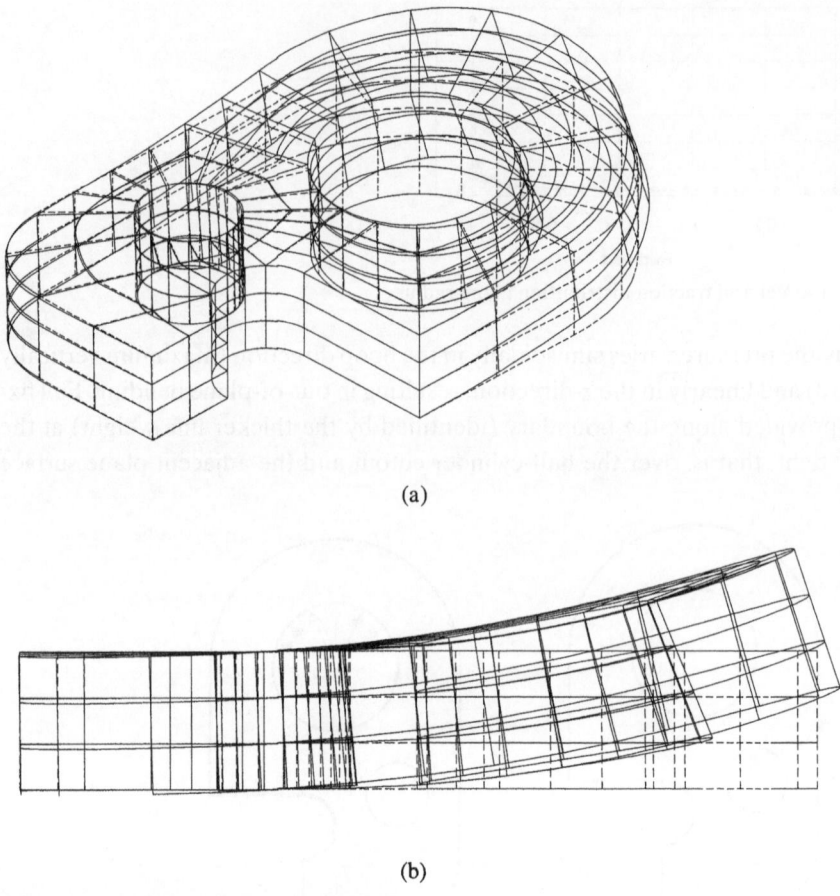

(a)

(b)

Figure 6.6: Deformed mesh (exaggerated).

above it, while all remaining boundaries, including the cylindrical cutout at center left, are free. The elastic parameters for the axle bearing are: Young's modulus $E = 200$ GPa and Poisson's ratio $v = 0.3$.

The surfaces of the component are discretized using a total of 392 quadratic boundary elements, defined by 1,174 boundary nodes, as shown in Fig. 6.5. The input and output data for this example may be found in the files COMP3D.DAT and COMP3D.OUT, respectively. Figure 6.6 shows the deformed mesh, with displacements magnified by a factor of 3. As expected, the component primarily suffers bending about the x-axis.

The contours of vertical stress σ_{yy} on the front face (at $z = 10$) are plotted in Fig. 6.7 and reveal compressive vertical stress concentrations at the crown of the bearing surface and tensile stresses at the sides. By contrast, under these loading conditions, the lower left half of the component is redundant. For design purposes, plots of equivalent stress would provide clearer indications of regions of overstress, and these can be readily determined from the output data.

Figure 6.8 is a plot of the vertical stress contours on the remaining surfaces. The symmetry about the $z = 0$ plane is evident. Although these data are unremarkable

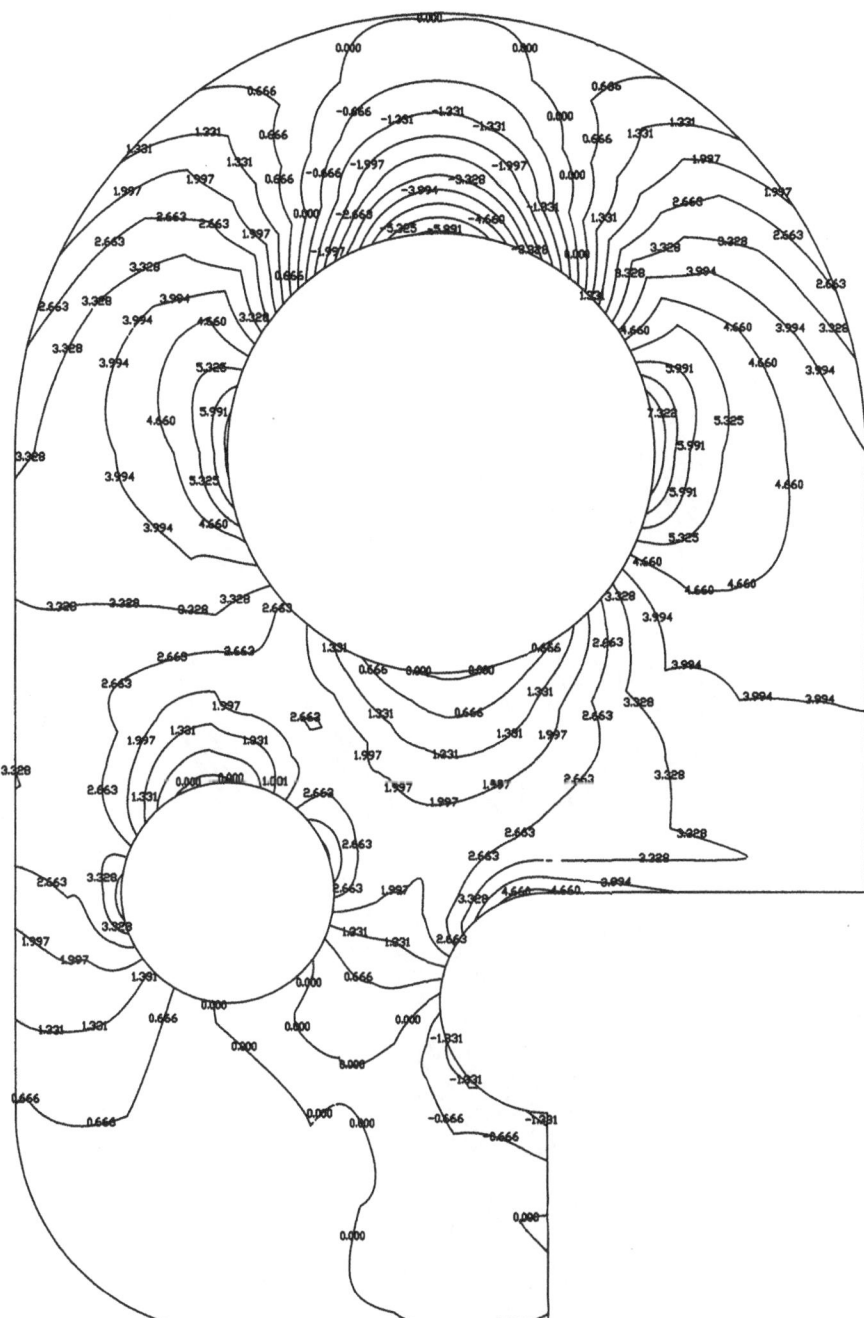

Figure 6.7: Contours of vertical stress σ_{yy} over front surface.

in themselves, it is worth noting that the contours are smooth, in contradistinction to the contour plots usually obtained from finite element methods. In that method, stresses are determined at the Gauss collocation points and must be extrapolated to the element boundaries (which include the region boundaries). Since equilibrium is not satisfied along finite element boundaries, averaging of

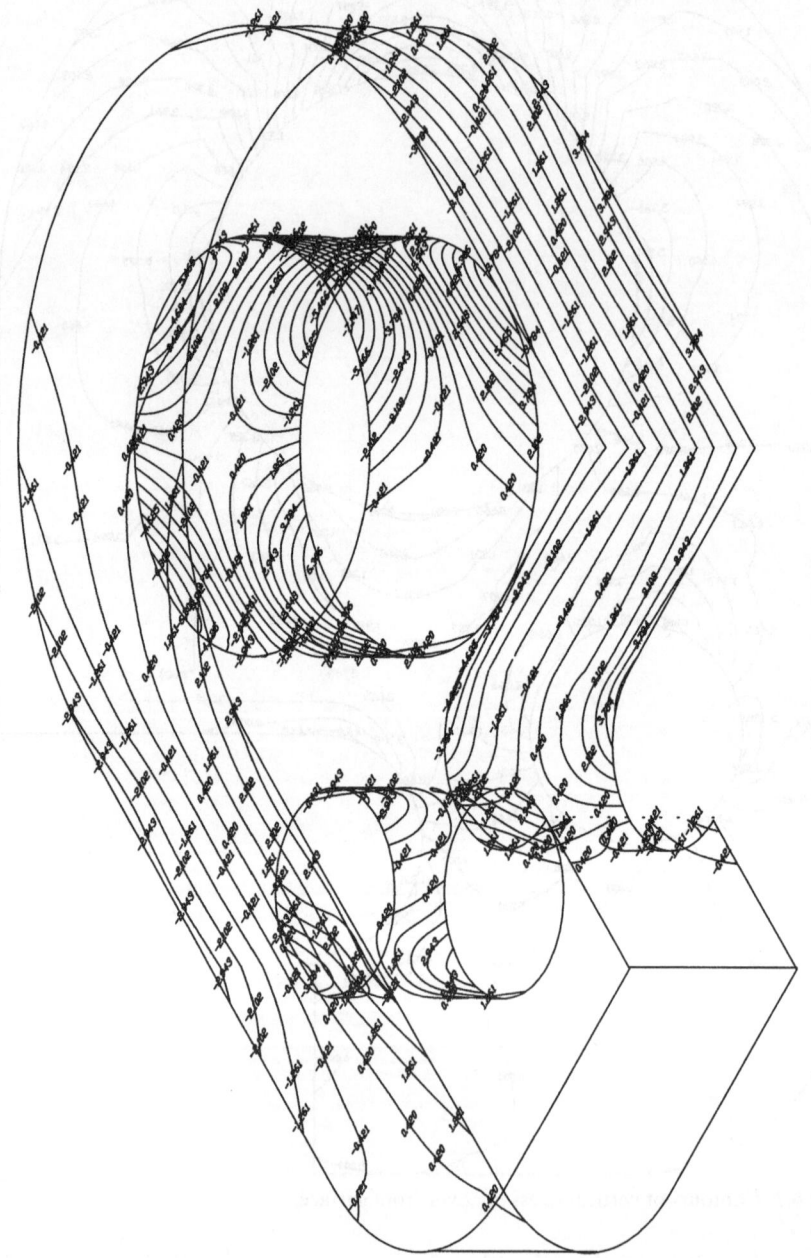

Figure 6.8: Contours of vertical stress σ_{yy} over (part of) component surface.

the extrapolated data at the interelement boundaries is essential. Sometimes, the stress discontinuities are so great that averaging is a leap of faith, whereas at the region boundaries averaging is obviously impossible. The consequences of this are that stress contours obtained using finite element analyses are often far inferior to those obtained using boundary element analyses.

6.5 Closure

The applications described here should be sufficient to demonstrate not only the high level of accuracy attainable by the boundary element method but also its practical efficiency in terms of data preparation and, less obviously perhaps, output interpretation. In particular, the modeling of the machine component illustrates the practical benefits of the "boundary only" discretization. Admittedly, a large number of boundary elements were employed, but a finite element analysis of comparable accuracy would have required far more elements, not only because one would have had to subdivide the interior into cells but also because finer surface discretization would have been required too. The finite element data would, of course, yield the displacements and stresses in the interior, but the advantages of this are generally illusory. More often than not, one is solely interested in boundary quantities only, and laborious filtering of the finite element output data becomes necessary. Naturally, using the boundary element method, one can determine internal quantities at selected points too (as demonstrated by the thick-cylinder example), although this requires some further calculation.

PART TWO

NONLINEAR PROBLEMS

CHAPTER SEVEN

Rate-Independent Plasticity Theory

7.1 Introduction

When solids are subjected to high stresses, irrecoverable (plastic) strains occur. For many materials, rate-independent elasto–plastic theory provides a useful framework for analyzing their behavior in this state. The theory is based on three main ideas: (a) the yield criterion (function), which defines the limits of elastic behavior; (b) the flow rule, which relates the irrecoverable strain rate to the yield functions (or plastic potentials), and (c) the hardening law, which determines the evolution of the yield function with irrecoverable deformation. In the classical (flow) theory of plasticity, the flow rule is based on Drucker's postulate (Drucker, 1959): this approach is commonly employed in numerical analysis (e.g., Owen & Hinton, 1980; Crisfield, 1997). However, Drucker's postulate is based on a nonnegative work assumption over a closed stress cycle, and is only suitable for stable materials – with perfect plasticity as a limiting case (Naghdi & Trapp, 1975a). Alternatively, Il'iushin's postulate (Il'iushin, 1961; Naghdi & Trapp 1975a,b), which is based on a nonnegative work assumption over a closed strain cycle, has no such limitation and can be used for both stable and unstable materials. To illustrate these ideas, we concentrate on four well-known yield criteria, namely, the Tresca, Von Mises, Mohr–Coulomb, and Drucker–Prager criteria. We then derive constitutive equations for hardening, softening, and perfect plastic materials governed by these criteria using isotropic hardening and kinematic hardening rules.

7.2 Isotropic Yield Criteria

A yield criterion is a function of stresses that separates elastic deformation states from elasto–plastic deformation states. If a yield criterion is plotted in stress space, it forms a surface (the yield surface) separating the two states. Within the yield surface, the deformation is purely elastic. Once the stress level reaches the yield surface, elasto–plastic deformations occur. On physical grounds, for isotropic materials, a yield criterion should be independent of the orientation of the coordinate system employed and therefore it should be a function of the stress invariants or, equivalently, the principal stresses σ_1, σ_2, and σ_3 only. Thus, before proceeding further, we need to review the definition of these quantities.

7.2.1 Stress Invariants and Principal Stresses

The traction–stress relationship (refer to Chapter Two) on a plane with unit outward normal n is

$$t_i = \sigma_{ij} n_j \tag{7.1}$$

Because the traction vector depends on the orientation of the plane, we can always find planes over which the tangential (shear) tractions vanish, leaving only the normal tractions. We call these planes (which are mutually orthogonal) "principal planes," and the corresponding normal tractions to these planes are called "principal stresses." The tractions over such a plane can be written as

$$t_i = \sigma \delta_{ij} n_j \tag{7.2}$$

where σ is the corresponding principal stress. Combining these two equations, we obtain

$$(\sigma_{ij} - \sigma \delta_{ij}) n_j = 0 \tag{7.3}$$

Not all the components of the outward normal can be zero; it follows that

$$\det(\sigma_{ij} - \sigma \delta_{ij}) = 0 \tag{7.4}$$

which yields, on expansion of the determinant, the cubic equation

$$\sigma^3 - I_1 \sigma^2 + I_2 \sigma - I_3 = 0 \tag{7.5}$$

where the constants I_1, I_2, and I_3 are

$$
\begin{aligned}
I_1 &= \sigma_{kk} \\
I_2 &= \frac{1}{2}(\sigma_{ij}\sigma_{ij} - \sigma_{ii}\sigma_{jj}) \\
I_3 &= \det \sigma_{ij}
\end{aligned}
\tag{7.6}
$$

These quantities are called the invariants of the stress tensor σ_{ij}, because it can be shown that they are independent of the orientation of the coordinate system. Of course, Eq. (7.5) has three roots, namely, σ_1, σ_2, and σ_3 (the principal stresses). Now, since this set of principal stresses is equivalent to σ_{ij}, it follows that

$$
\begin{aligned}
I_1 &= \sigma_1 + \sigma_2 + \sigma_3 \\
I_2 &= -(\sigma_1\sigma_2 + \sigma_2\sigma_3 + \sigma_3\sigma_1) \\
I_3 &= \sigma_1\sigma_2\sigma_3
\end{aligned}
\tag{7.7}
$$

In addition, we can define deviatoric stresses as follows:

$$\sigma'_{ij} = \sigma_{ij} - \frac{1}{3}\sigma_{kk}\delta_{ij} \tag{7.8}$$

These stresses are closely linked to shear stresses, which play an important role in yield. They too can be manipulated to form invariants (of the deviatoric stress tensor), giving

$$J_1 = \sigma'_{kk} = 0$$

$$J_2 = \frac{1}{2}\sigma'_{ij}\sigma'_{ij} = \frac{1}{6}[(\sigma_1 - \sigma_2)^2 + (\sigma_2 - \sigma_3)^2 + (\sigma_3 - \sigma_1)^2] \tag{7.9}$$

$$J_3 = \frac{1}{3}\sigma'_{ij}\sigma'_{jk}\sigma'_{ki}$$

Finally, the principal stresses can be expressed (Owen & Hinton, 1980) in terms of the stress invariants as follows:

$$\sigma_M = \frac{2}{\sqrt{3}}\sqrt{J_2}\sin(\theta + \beta_M) + \frac{1}{3}I_1 \tag{7.10}$$

where $M = 1, 2, 3$, $\beta_1 = 2\pi/3$, $\beta_2 = 0$, $\beta_3 = 4\pi/3$, $-\pi/6 \le \theta \le \pi/6$, and

$$\theta = \frac{1}{3}\sin^{-1}\left[-\frac{3\sqrt{3}}{2}\frac{J_3}{J_2^{3/2}}\right] \tag{7.11}$$

These results will be useful in the following subsections, which contain descriptions of four common yield criteria, and in the remainder of this chapter.

7.2.2 The Tresca Criterion

The Tresca yield criterion (1864) predates all the others. According to this criterion, plastic deformation occurs when the maximum shear stress attains a critical value k. If we assume that $\sigma_1 \ge \sigma_2 \ge \sigma_3$, then this criterion may be expressed as

$$\frac{\sigma_1 - \sigma_3}{2} = k(h^\alpha) \tag{7.12}$$

where k is some function of the hardening parameters $h^\alpha(\alpha = 1, 2, \ldots)$. These hardening parameters (also called internal variables) must be determined experimentally. They express the relationship between the elasto–plastic loading history (often simply characterized by the accumulated plastic strain) and the critical shear stress. In perfectly plastic materials, k is a constant. The Tresca yield surface in principal stress space is shown as in Fig. 7.1a and the projection of this surface on the so-called π-plane, that is, the plane normal to the space diagonal, is shown in Fig. 7.1b.

From Fig. 7.1a, we see that the Tresca criterion is the surface of an infinitely long regular hexagonal cylinder. This implies that hydrostatic stress does not cause yielding. Hence, this criterion is appropriate for materials such as metals (and undrained clays, in a "total" stress analysis).

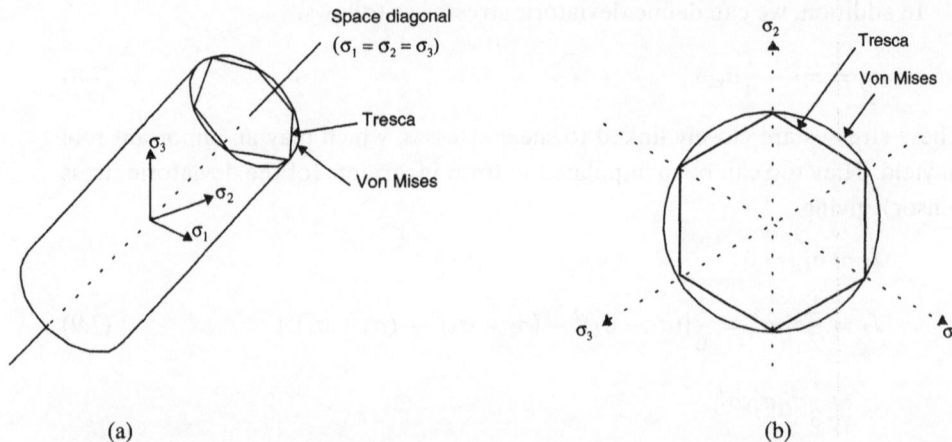

Figure 7.1: Tresca and Von Mises yield surfaces in (a) principal stress space and (b) the π-plane.

7.2.3 The Von Mises Criterion

An alternative to the Tresca criterion, suggested by Von Mises (1913), has the form

$$\sqrt{J_2} = k(h^\alpha) \tag{7.13}$$

where k is a material parameter and J_2 is the second invariant of the deviatoric stress tensor. As shown in Fig. 7.1, this criterion takes the form of a cylinder oriented along the space diagonal. Again, hydrostatic stress has no influence on yielding.

7.2.4 The Mohr–Coulomb Criterion

For soils, rocks, and concrete, changes in mean stress can cause yielding, and a different type of criterion is necessary for these materials. One such criterion stems from Coulomb's equation (1773), which is essentially a statement of equilibrium for a frictional material. Coulomb's equation relates the normal and shear stresses on a failure plane:

$$\tau = c - \sigma_n \tan \phi \tag{7.14}$$

where τ is the shear stress, σ_n is the normal stress (note that tensile stress is positive), c is the cohesion, and ϕ is the angle of internal friction. Both c and ϕ are experimentally determined material constants. This criterion may be written (for $\sigma_1 \geq \sigma_2 \geq \sigma_3$) as

$$\frac{\sigma_1 - \sigma_3}{2} = c \cos \phi - \frac{\sigma_1 + \sigma_3}{2} \sin \phi \tag{7.15}$$

In principal stress space, this criterion describes an irregular hexagonal cone, as shown in Fig. 7.2. For frictionless materials ($\phi = 0$), the Mohr–Coulomb criterion reduces to the Tresca criterion.

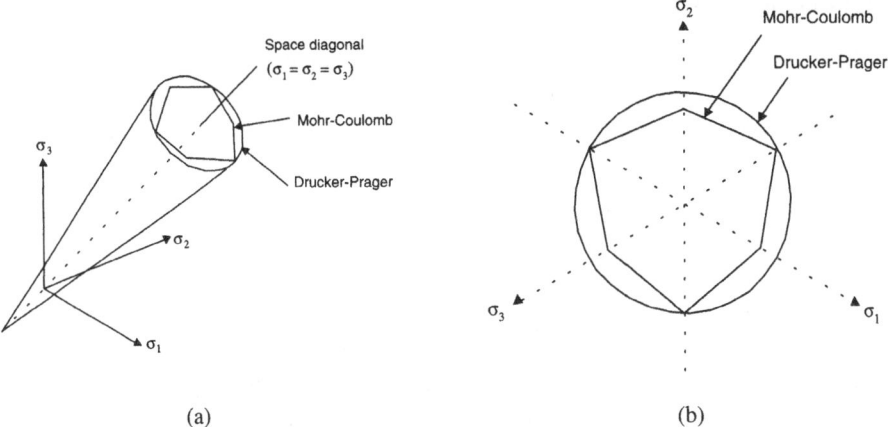

Figure 7.2: Mohr–Coulomb and Drucker–Prager yield surfaces in (a) principal stress space and (b) the π-plane.

7.2.5 The Drucker–Prager Criterion

Another yield criterion that is mean-stress dependent was proposed by Drucker & Prager (1952). This criterion combines Coulomb friction with the Von Mises yield criterion to give

$$\alpha I_1 + \sqrt{J_2} = k \tag{7.16}$$

where α and k are material constants. This criterion describes a cone in principal stress space, as shown in Fig. 7.2. By matching the dimensions of the Mohr–Coulomb and Drucker–Prager surfaces in stress space, we can obtain explicit relationships between their parameters. Thus, if we match the Drucker–Prager cone with the outer vertices of the Mohr–Coulomb surface, we obtain

$$\alpha = \frac{2\sin\phi}{\sqrt{3}(3 - \sin\phi)}, \qquad k = \frac{6c\cos\phi}{\sqrt{3}(3 - \sin\phi)} \tag{7.17}$$

Alternatively, coincidence with the inner vertices of the Mohr–Coulomb hexagon results from employing the relationships

$$\alpha = \frac{2\sin\phi}{\sqrt{3}(3 + \sin\phi)}, \qquad k = \frac{6c\cos\phi}{\sqrt{3}(3 + \sin\phi)} \tag{7.18}$$

Because of their simplicity, the Mohr–Coulomb and Drucker–Prager yield criteria have been used extensively to model the behavior of geotechnical materials, despite the fact that they predict unrealistic volume dilation, in "associated" plasticity formulations. Further, the Drucker–Prager yield criterion evidently deviates significantly from the Mohr–Coulomb criterion in some parts of stress space. For more advanced models, the reader should consult the specialist literature.

7.3 Principles of Elasto–Plastic Flow

In this section, we briefly review some of basic principles of elasto–plastic flow. Although the theory is difficult and, in parts, still controversial, we cannot delve very far into these matters here. This sometimes means that the reader is invited to accept statements on trust, but fuller accounts of the theory are readily available in the references cited. Moreover, to preserve continuity, some mathematical manipulations (which are original in part) have been relegated to Appendix D. In this development of the theory, we employ the more rigorous Il'iushin strain-space formulation, rather than the Drucker stress-space formulation. Inevitably, much of this chapter will be difficult for those new to plasticity theory, and so we provide an introduction to the subject in Appendix G. This appendix may also be worth reviewing, for its description of the Drucker stress-space formulation, by readers with some knowledge of plasticity theory.

Turning now to the Il'iushin theory, we begin by rewriting the yield functions described earlier in the general form

$$f(\sigma, \varepsilon^p, h^\alpha) = 0, \tag{7.19}$$

where σ and ε^p are the stress and plastic strain tensors, respectively. The internal variables h^α are normally functions of the irrecoverable deformation history, characterized by equivalent plastic strain, plastic work, etc. If we wish to take nonlinear elastic deformation into account and allow for the fact that material properties vary with irrecoverable deformation, the total stress–strain relationships may be written in the form

$$\sigma = \hat{\sigma}(\varepsilon, \varepsilon^p, h^\alpha) \tag{7.20}$$

We now assume that the total stress tensor σ is derivable from a potential, such that the partial derivative of σ with respect to ε is symmetric (Naghdi & Trapp, 1975b), that is,

$$\frac{\partial \hat{\sigma}_{ij}}{\partial \varepsilon_{kl}} = \frac{\partial \hat{\sigma}_{kl}}{\partial \varepsilon_{ij}} \quad \text{or} \quad D_{ijkl} = D_{klij} \tag{7.21}$$

where

$$D_{ijkl} = \frac{\partial \hat{\sigma}_{ij}}{\partial \varepsilon_{kl}} \tag{7.22}$$

Now, the yield function (Eq. 7.19), also called the loading function, is generally formulated in stress space: it is couched in terms of stresses rather than strains. We now assume that there is a loading function $g(\varepsilon, \varepsilon^p, h^\alpha) = 0$ in strain space that is equivalent to the loading function in stress space such that

$$g(\varepsilon, \varepsilon^p, h^\alpha) = f(\hat{\sigma}(\varepsilon, \varepsilon^p, h^\alpha), \varepsilon^p, h^\alpha) = 0 \tag{7.23}$$

The Il'iushin postulate states that the work done in a closed strain cycle is nonnegative, that is,

$$\oint \sigma : d\varepsilon = \int_{t_1}^{t_2} \sigma : \dot{\varepsilon} dt \geq 0 \qquad (7.24)$$

in which t_1 and t_2 represent the beginning and ending times of the strain cycle, the superposed dot denotes the derivative with respect to time, and the colon (:) indicates the scalar (i.e., inner) product of the two tensors. This postulate, which is valid for hardening, softening, and perfectly plastic materials, can be used to derive quite general results in strain space. In particular, one can derive the "plastic flow rule" (as described in some detail in Appendix D.1):

$$\dot{\sigma} - \mathbf{D}^e : \dot{\varepsilon} = -\dot{\lambda} \frac{\partial g}{\partial \varepsilon} \qquad (7.25)$$

where $\dot{\lambda}$ is a nonnegative scaling factor called the plastic multiplier and $\mathbf{D}^e$ is the elastic constitutive tensor (i.e., D_{ijkl} within the yield surface). The plastic flow rule provides the basis for the development of the elasto–plastic stress–strain relationships. In addition, the inequality (the "plastic loading rule")

$$\frac{\partial g}{\partial \varepsilon} : \dot{\varepsilon} \geq 0 \qquad (7.26)$$

arises naturally from the Il'iushin stability postulate and can be used to establish the validity of the deformation path.

However, it is difficult to use these equations directly because the loading function is normally expressed in stress space, not strain space. Consequently, we need to translate them into their stress–space equivalents by employing the relationships (Casey & Naghdi 1981)

$$\frac{\partial g}{\partial \varepsilon} = \mathbf{D}^e : \frac{\partial f}{\partial \sigma}$$

$$\frac{\partial g}{\partial \varepsilon^p} = \frac{\partial f}{\partial \varepsilon^p} + \frac{\partial f}{\partial \sigma} : \frac{\partial \hat{\sigma}}{\partial \varepsilon^p} \qquad (7.27)$$

$$\frac{\partial g}{\partial h^\alpha} = \frac{\partial f}{\partial h^\alpha} + \frac{\partial f}{\partial \sigma} : \frac{\partial \hat{\sigma}}{\partial h^\alpha}$$

Substitution of these equations leads to the analogous results in stress space:

$$\dot{\sigma} = \mathbf{D}^e : \dot{\varepsilon} - \dot{\lambda} \mathbf{D}^e : \frac{\partial f}{\partial \sigma} \qquad (7.28)$$

and

$$\frac{\partial f}{\partial \sigma} : \mathbf{D}^e : \dot{\varepsilon} \geq 0 \qquad (7.29)$$

The first of these equations is immediately recognizable as a stress–strain relationship, cast in incremental form. The irrecoverable (plastic) strains are given by the term $\dot{\lambda} (\partial f / \partial \sigma)$, where $\dot{\lambda}$ is a yet undetermined parameter (which defines the magnitude of the resultant plastic strain) and $\partial f / \partial \sigma$ signifies the outward normal to the yield surface. Thus, the resultant of the plastic strains is normal to the yield surface (normality). Although Naghdi & Trapp (1975b) developed the strain-space theory for finite elasto–plastic deformation, in the following we only consider its applications to infinitesimal elasto–plastic deformation.

7.4 Constitutive Relationships

In elasto–plastic deformation, the total strain increment can be decomposed into elastic and plastic parts as follows:

$$\dot{\varepsilon} = \dot{\varepsilon}^e + \dot{\varepsilon}^p \tag{7.30}$$

The incremental stress–strain response can then be written in the form

$$\dot{\sigma} = \mathbf{D}^e : \dot{\varepsilon}^e = \mathbf{D}^e : (\dot{\varepsilon} - \dot{\varepsilon}^p) \tag{7.31}$$

where $\mathbf{D}^e$ is the elastic constitutive tensor, derived from the generalized Hooke's law, as described in Chapter 2. Comparing this equation with the general stress–strain equation derived earlier (Eq. 7.28) confirms our observation that the plastic strain increments are defined by the expression

$$\dot{\varepsilon}^p = \lambda \frac{\partial f}{\partial \sigma} \tag{7.32}$$

This equation defines the relationship between the components of the plastic strain increments: their resultant is normal to the yield surface. In a physical sense, it defines how the material flows plastically; hence, the terms "flow rule" and "normality." During plastic loading, the yield surface may shift, expand, or change shape because of irrecoverable deformation. However, the strain (and stress) state must always remain on the yield surface, because states outside a yield surface are inadmissible whereas states inside the surface are elastic. This means that the yield function is always satisfied during plastic loading and implies the "consistency condition" in stress space:

$$\dot{f}(\sigma, \varepsilon^p, h^\alpha) = 0 \tag{7.33}$$

and its equivalent in strain space:

$$\dot{g}(\varepsilon, \varepsilon^p, h^\alpha) = 0 \tag{7.34}$$

Expanding the strain-space consistency condition produces

$$\frac{\partial g}{\partial \varepsilon} : \dot{\varepsilon} + \frac{\partial g}{\partial \varepsilon^p} : \dot{\varepsilon}^p + \frac{\partial g}{\partial h^\alpha} \dot{h}^\alpha = 0 \tag{7.35}$$

For convenience, we now express the increment of the internal variable as a function of the plastic multiplier:

$$h^\alpha = h_{\alpha,\lambda} \lambda \tag{7.36}$$

where $h_{\alpha,\lambda} = \partial h^\alpha / \partial \lambda$ can be determined for specific yield functions, as will be demonstrated later. Combining Eqs. (7.32), (7.35), and (7.36), we can rewrite the plastic multiplier in the form

$$\dot{\lambda} = \frac{1}{\psi} \frac{\partial g}{\partial \varepsilon} : \dot{\varepsilon} \tag{7.37}$$

where

$$\psi = -\left(\frac{\partial g}{\partial \varepsilon^p} : \frac{\partial f}{\partial \sigma} + \frac{\partial g}{\partial h^\alpha} h_{\alpha,\lambda} \right) \tag{7.38}$$

We now need to transform these equations into stress space. To do this, first we note that the general stress–strain equation (7.20) and its specific form (7.31) provide the simplifications

$$\frac{\partial \hat{\sigma}}{\partial \varepsilon^p} = -\mathbf{D}^e$$
$$\frac{\partial \hat{\sigma}}{\partial h^\alpha} = 0 \tag{7.39}$$

Substituting these relationships into the transformation equations (7.27) produces the desired result in stress space:

$$\dot{\lambda} = \frac{1}{\psi} \frac{\partial f}{\partial \sigma} : \mathbf{D}^e : \dot{\varepsilon} \tag{7.40}$$

where

$$\psi = \frac{\partial f}{\partial \sigma} : \mathbf{D}^e : \frac{\partial f}{\partial \sigma} - \frac{\partial f}{\partial \varepsilon^p} : \frac{\partial f}{\partial \sigma} - \frac{\partial f}{\partial h^\alpha} h_{\alpha,\lambda} \tag{7.41}$$

We now have an explicit expression for the plastic multiplier and this can be substituted into the stress–strain relationship (Eq. 7.28) to obtain

$$\dot{\sigma} = \left(\mathbf{D}^e - \frac{\langle \hat{g} \rangle}{\psi} \mathbf{D}^e : \frac{\partial f}{\partial \sigma} \otimes \frac{\partial f}{\partial \sigma} : \mathbf{D}^e \right) : \dot{\varepsilon} \tag{7.42}$$

where

$$\hat{g} = \frac{\partial f}{\partial \sigma} : \mathbf{D}^e : \dot{\varepsilon} \tag{7.43}$$

and $\otimes$ indicates the indeterminate vector product (i.e., the "outer" product). The parameter $\hat{g}$ is used to determine the magnitude of $\langle \hat{g} \rangle$, according to the loading rule (Eq. 7.29), which in effect results in the following scenarios:

$$\langle \hat{g} \rangle = \begin{cases} 0 & \text{if } f(\sigma, \varepsilon^p, h^\alpha) < 0 & (\rightarrow \text{elastic}) \\ 1 & \text{if } f(\sigma, \varepsilon^p, h^\alpha) = 0 \quad \text{and} \quad \hat{g} \geq 0 & (\rightarrow \text{loading}) \\ 0 & \text{if } f(\sigma, \varepsilon^p, h^\alpha) = 0 \quad \text{and} \quad \hat{g} < 0 & (\rightarrow \text{unloading}) \end{cases} \tag{7.44}$$

During loading ($f = 0$, $\hat{g} > 0$), a material may be in hardening, softening, or ideally (perfect) plastic states. Following Casey & Naghdi (1981), we introduce the rate-independent quantity Γ to distinguish between the three deformation states (see Appendix D.4):

$$\Gamma > 0 \quad \text{(hardening)}$$
$$\Gamma = 0 \quad \text{(ideal plasticity)} \qquad (7.45)$$
$$\Gamma < 0 \quad \text{(softening)}$$

where

$$\Gamma = -\frac{\partial f}{\partial \varepsilon^p} : \frac{\partial f}{\partial \sigma} - \frac{\partial f}{\partial h^\alpha} h_{\alpha,\lambda} \qquad (7.46)$$

To apply these results (specifically Eq. 7.42) in practice, it is first necessary to calculate the derivative $\partial f/\partial \sigma$ and the parameter ψ, which is itself a function of various derivatives of the yield function. These calculations for the four yield functions described earlier are the subject of the following sections.

7.5 Isotropic Hardening Materials

In isotropic hardening materials, the yield surface is assumed to expand (or shrink) uniformly in stress space during plastic loading, as shown in Fig. 7.3. This is a

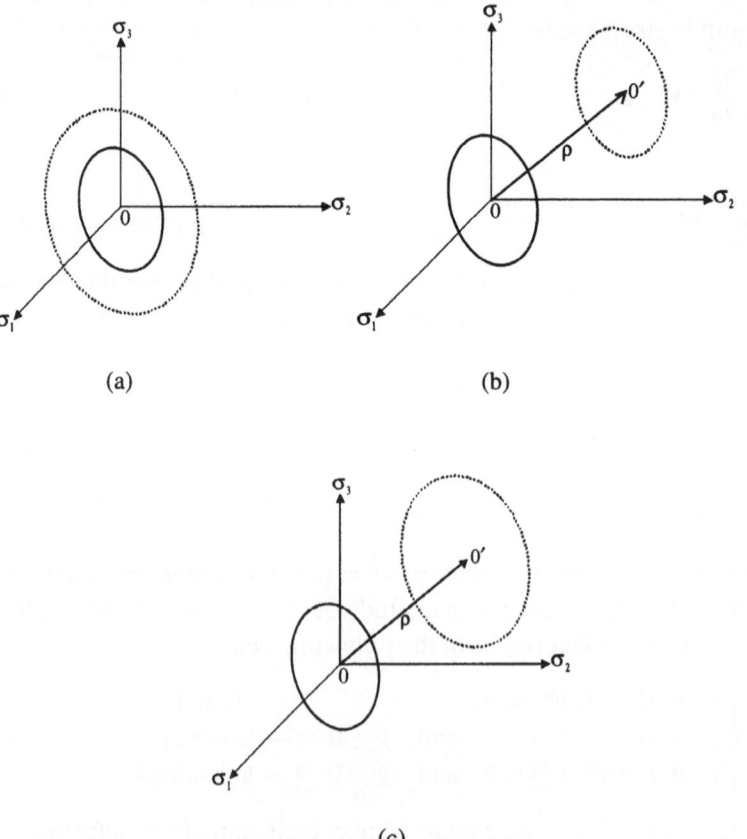

Figure 7.3: Evolution of yield functions for (a) isotropic hardening, (b) kinematic hardening, and (c) mixed hardening.

convenient simplification of real material behavior, but one that is often suffi-ciently accurate for cases where loading increases monotonically. However, for more complex stress paths or load reversals (cyclic loading), this model is gener-ally inadequate.

To apply this model of material behavior, it is first necessary to carry out load tests to determine the material constants. Often, simple uniaxial loading tests are performed and therefore it is necessary to derive the uniaxial forms of the yield functions to establish equivalence between the constants for uniaxial and general states of stress. The way in which yield stresses change with plastic deformation also requires that we establish equivalences between measures of plastic deformation in uniaxial and general states of stress. These matters are the subjects of the following two subsections. In the last subsection, we finally establish the explicit results required to evaluate the derivatives of the yield functions and the parameter ψ.

7.5.1 Equivalent Uniaxial Yield Criteria

In general, the state variables (stresses) and the material parameters in yield functions can be separated, which allows us to write

$$f(\sigma, \varepsilon^p, h^\alpha) = \bar{f}(\sigma) - \sigma_y(h^\alpha) = 0 \tag{7.47}$$

where $\sigma_y(h^\alpha)$ is the uniaxial yield stress (which may be a function of plastic strain) and $\bar{f}(\sigma)$ is a function of the current state of stress only. Of interest here is the reduced form of $\bar{f}(\sigma)$ for the special case of a uniaxial stress state. By definition, in the uniaxial stress state, all stress components are zero, except one ($\sigma_{11} = -\sigma_{xx}$, say). Under these circumstances, the stress invariants reduce to the simple forms

$$
\begin{aligned}
I_1 &= -\sigma_{xx} \\
J_2 &= (\sigma_{xx})^2/3 \\
J_3 &= -2(\sigma_{xx}/3)^3 \\
\theta &= \pi/6
\end{aligned}
\tag{7.48}
$$

If we let $\sigma_y(0)$ be the uniaxial stress at first yield, then the reduced forms for the four yield criteria discussed previously may be readily determined, using Eq. (7.10). The results of these calculations are listed in Table 7.1, in which the constants γ and χ are defined as follows: $\gamma = (1 - \sin\phi)/2$, $\chi = -\alpha + 1/\sqrt{3}$; the Drucker–Prager constants α and k are determined from either Eq. (7.17) or Eq. (7.18).

7.5.2 Equivalent Plastic Strain

The yield stress $\sigma_y(h^\alpha)$ may increase (hardening), decrease (softening), or remain constant (perfect plasticity) during plastic deformation. Various internal variables, such as plastic work, can be used to characterize this history of plastic deformation. Here, we assume that some measure of plastic strain, denoted by $\bar{\varepsilon}^p$, fulfills this

Table 7.1. Equivalent uniaxial yield criteria

Criterion	Yield Function	Yield Stress
Tresca	$2\sqrt{J_2}\cos\theta$	$\sigma_y(0)$
Von Mises	$\sqrt{3J_2}$	$\sigma_y(0)$
Mohr–Coulomb	$\dfrac{1}{\gamma}\left[\dfrac{\sin\phi}{3}I_1 + \sqrt{J_2}\left(\cos\theta - \dfrac{\sin\phi}{\sqrt{3}}\sin\theta\right)\right]$	$\dfrac{1}{\gamma}c\cos\phi$
Drucker–Prager	$\dfrac{1}{\chi}(\alpha I_1 + \sqrt{J_2})$	$\dfrac{1}{\chi}k$

role, that is, $h^\alpha = \bar\varepsilon^p$. One sensible definition of the plastic strain increment is

$$\dot{\bar\varepsilon}^p = c'\sqrt{\dot\varepsilon^p : \dot\varepsilon^p} = h_{\varepsilon,\lambda}\dot\lambda \tag{7.49}$$

in which Eq. (7.32) has been employed to link it with the plastic multiplier via the parameter $h_{\varepsilon,\lambda}$, where, by definition,

$$h_{\varepsilon,\lambda} = c'\sqrt{\frac{\partial f}{\partial\sigma} : \frac{\partial f}{\partial\sigma}} \tag{7.50}$$

The arbitrary constant c' in the above equations is now assigned a magnitude such that the plastic strain increment defined by Eq. (7.49) is equal to the plastic strain increment produced by an uniaxial stress test. In other words, the parameter c' is chosen such that Eq. (7.49) produces an "equivalent" plastic strain increment. It can be shown (Chen, 1994) that this amounts to calculating the value of c' from the equation

$$c' = \frac{\partial f}{\partial\sigma_{11}} \Bigg/ \sqrt{\frac{\partial f}{\partial\sigma_{ij}}\frac{\partial f}{\partial\sigma_{ij}}} \tag{7.51}$$

The results of these calculations, for four yield criteria, are listed in Table 7.2.

7.5.3 Explicit Derivations

For a material in which the yield stress $\sigma_y(h^\alpha)$ is a function of the equivalent plastic strain (i.e., $h^\alpha = \bar\varepsilon^p$), the hardening rule can be written in the form

$$\sigma_y(h^\alpha) = \sigma_y(0) + \int H'd\bar\varepsilon^p \tag{7.52}$$

Table 7.2. Magnitude of the multiplier c' for four yield criteria

Tresca	Von Mises	Mohr–Coulomb	Drucker–Prager
$\sqrt{\dfrac{2}{3}}$	$\sqrt{\dfrac{2}{3}}$	$\dfrac{\sqrt{2}(1-\sin\phi)}{\sqrt{3-2\sin\phi+3\sin^2\phi}}$	$\dfrac{1/\sqrt{3}-\alpha}{\sqrt{3\alpha^2+1/2}}$

where H' is the slope of the uniaxial stress–plastic strain curve. This parameter can be determined from experiment. To establish the constitutive relationships in concrete form, we now need to determine the explicit forms of the function ψ and the derivatives of the yield functions $\partial f/\partial \sigma$, which appear in Eq. (7.42). Using the special form of the yield functions defined in Eq. (7.47) and the hardening rule defined above (Eq. 7.52), we can readily establish the following derivatives:

$$\frac{\partial f}{\partial \sigma} = \frac{\partial \bar{f}}{\partial \sigma}$$

$$\frac{\partial f}{\partial \varepsilon^p} = 0 \qquad (7.53)$$

$$\frac{\partial f}{\partial h^\alpha} = -\frac{\partial \sigma_y(h^\alpha)}{\partial \bar{\varepsilon}^p} = -H'$$

After substituting these equations into Eq. (7.41), and recalling that $\bar{\varepsilon}^p$ is taken to be the internal variable (that is, $h_{\alpha,\lambda} = h_{\varepsilon,\lambda}$), the parameter ψ simplifies to

$$\psi = \frac{\partial \bar{f}}{\partial \sigma} : \mathbf{D}^e : \frac{\partial \bar{f}}{\partial \sigma} + H' h_{\varepsilon,\lambda} \qquad (7.54)$$

To calculate this parameter, all that now remains is to determine the derivative of the yield function, since all other parameters either can be determined experimentally or (as in the case of the term $h_{\varepsilon,\lambda}$) are functions of this derivative. To calculate this quantity, we first rewrite it in terms of the stress invariants:

$$\frac{\partial \bar{f}}{\partial \sigma} = C_1 \frac{\partial I_1}{\partial \sigma} + C_2 \frac{\partial \sqrt{J_2}}{\partial \sigma} + C_3 \frac{\partial J_3}{\partial \sigma} \qquad (7.55)$$

where the coefficients C_1, C_2, and C_3 are, of course,

$$C_1 = \frac{\partial \bar{f}}{\partial I_1}$$

$$C_2 = \frac{\partial \bar{f}}{\partial \sqrt{J_2}} \qquad (7.56)$$

$$C_3 = \frac{\partial \bar{f}}{\partial J_3}$$

These coefficients will be different for different yield functions. Now, by using Eq. (7.11), the last of these becomes

$$\begin{aligned} C_3 &= \frac{\partial \bar{f}}{\partial \theta} \frac{\partial \theta}{\partial J_3} \\ &= -\frac{\sqrt{3}}{2 J_2^{3/2} \cos 3\theta} \frac{\partial \bar{f}}{\partial \theta} \end{aligned} \qquad (7.57)$$

Explicit expressions for the three constants C_1–C_3 are listed in Table 7.3, for four different yield criteria.

Finally, to calculate the derivative of the yield functions, we also need to calculate the derivatives of the stress invariants indicated in Eq. (7.55). This is a

Table 7.3. Yield criteria constants, C_1–C_3 (for $\theta \neq \pm \pi/6$)

Yield Criterion	C_1	C_2	C_3
Tresca	0	$2\cos\theta(1 + \tan\theta \tan 3\theta)$	$\dfrac{\sqrt{3}}{J_2}\dfrac{\sin\theta}{\cos 3\theta}$
Von Mises	0	$\sqrt{3}$	0
Mohr–Coulomb	$\dfrac{1}{3\gamma}\sin\phi$	$\dfrac{\cos\theta}{\gamma}[1 + \tan\theta \tan 3\theta$ $+ \sin\phi(\tan 3\theta - \tan\theta)/\sqrt{3}]$	$\dfrac{\sqrt{3}\sin\theta + \cos\theta \sin\phi}{2\gamma J_2 \cos 3\theta}$
Drucker–Prager	$\dfrac{\alpha}{\chi}$	$\dfrac{1}{\chi}$	0

straightforward calculation, using the definitions given in Section 7.2.1. For convenience, we list the results here, namely,

$$\frac{\partial I_1}{\partial \sigma_{ij}} = \delta_{ij}$$

$$\frac{\partial \sqrt{J_2}}{\partial \sigma_{ij}} = \frac{\sigma'_{ij}}{2\sqrt{J_2}} \tag{7.58}$$

$$\frac{\partial J_3}{\partial \sigma_{ij}} = \sigma'_{ik}\sigma'_{kj} - \frac{2}{3}J_2\delta_{ij}$$

It should be noted that as θ approaches $\pm\pi/6$, the expressions for C_2 and C_3 (in Table 7.3) for the Tresca and Mohr–Coulomb criteria become indeterminate. These constants can, however, be derived directly from Table 7.1 (by setting $\theta = \pm\pi/6$), which yields the results shown in Table 7.4.

In the numerical implementation, when $|\theta| \leq 29°$, Table 7.3 (otherwise Table 7.4) is used to calculate the values of C_1, C_2, and C_3 for the Tresca and Mohr–Coulomb criteria. Physically, this artifice corresponds to a rounding off of the yield surface corners.

7.6 Kinematic Hardening Materials

Although isotropic hardening is a convenient assumption, in reality, yield surfaces translate (as well as expand or contract) as a consequence of accumulated plastic strain. Pure translation of a yield surface is called kinematic hardening, and

Table 7.4. Yield criteria constants, C_1–C_3 (for $\theta = \pm \pi/6$)

Yield Criteria	C_1	C_2	C_3
Tresca	0	$\sqrt{3}$	0
Mohr–Coulomb	$\dfrac{1}{3\gamma}\sin\phi$	$\dfrac{\sqrt{3}}{2\gamma} - \dfrac{\sin\theta \sin\phi}{\sqrt{3}\gamma}$	0

among other things, it gives rise to such phenomena as the Bauschinger effect (Hill, 1950) when materials are subjected to cyclic loading. Kinematic hardening yield functions have the form

$$\bar{f}(\sigma - \rho) - \sigma_y = 0 \qquad (7.59)$$

where σ_y is the (constant) yield stress limit and ρ is the center of the yield surface and is called the back stress. Figure 7.3b illustrates kinematic hardening in stress space. Various theories have been advanced to describe how yield surfaces translate during kinematic hardening. Here, we adopt the hardening rule (Ziegler, 1959), in which the incremental change in back stress is assumed to be

$$\dot{\rho} = \dot{\mu}(\sigma - \rho) \qquad (7.60)$$

where $\dot{\mu}$ is a positive proportionality factor. This means that the back stress translates towards the current stress. We now assume that the proportionality factor is a function of the equivalent plastic strain $\bar{\varepsilon}^p$, that is,

$$\dot{\mu} = \frac{\partial \mu}{\partial \bar{\varepsilon}^p} \dot{\bar{\varepsilon}}^p \qquad (7.61)$$

With these assumptions, it can be shown (see Appendix D.2) that

$$\frac{\partial \mu}{\partial \bar{\varepsilon}^p} = \frac{\bar{H}}{\sigma_y} \qquad (7.62)$$

where H is the slope of the uniaxial stress–plastic strain curve for a purely kinematic hardening material. Taking the last three equations together allows us to update the back stress during the incremental solution process. Doing some further manipulations (described in Appendix D.2), analogous to those described earlier for isotropic hardening, yields the equation

$$\psi = \frac{\partial \bar{f}}{\partial \sigma} : \mathbf{D}^e : \frac{\partial \bar{f}}{\partial \sigma} + \bar{H} h_{\varepsilon,\lambda} \qquad (7.63)$$

where $h_{\varepsilon,\lambda}$ is determined from Eq. (7.50). Because the expressions for the yield functions for isotropic hardening and kinematic hardening (Eqs. 7.47 and 7.59, respectively) are identical in form, all the results obtained for isotropic hardening also apply here, with the proviso that the stress tensor σ is replaced by the reduced stress tensor, $\sigma - \rho$.

7.7 Mixed Hardening Materials

In real materials, yield surfaces both translate and change in size during elasto–plastic deformation: this is termed mixed hardening and is illustrated in Fig. 7.3c. Although this type of hardening offers a more realistic model of material behavior than isotropic hardening or kinematic hardening alone, it still falls short of a complete description. For example, significant distortions and rotations of yield surfaces are evident in experimental data obtained from tests on soils. Evidently, there comes a point where utilization of more complex hardening models leads to diminishing returns, and we choose not to go further down this route. One advantage of the simple mixed hardening model is that the isotropic and kinematic components can be isolated. Thus, we begin by assuming that the plastic strain increment can be decomposed into isotropic and kinematic parts as follows:

$$
\begin{aligned}
\dot{\varepsilon}^p &= \dot{\varepsilon}^{pi} + \dot{\varepsilon}^{pk} \\
&= m\dot{\varepsilon}^p + (1-m)\dot{\varepsilon}^p
\end{aligned}
\tag{7.64}
$$

where m is a scale-factor ($0 \leq m \leq 1$) that signifies the proportion of the irrecoverable deformation that can be attributed to isotropic hardening. This parameter can be treated as a material parameter, or possibly as a state-dependent variable. Since translation and expansion are independent, the yield function is simply

$$
\bar{f}(\sigma - \rho) - \sigma_y(\bar{\varepsilon}^{pi}) = 0
\tag{7.65}
$$

where $\bar{\varepsilon}^{pi}$ is the equivalent plastic strain attributable to isotropic hardening. The hardening rule, for the expansion of the yield surface (compare Eq. 7.52) is

$$
\begin{aligned}
\sigma_y(\bar{\varepsilon}^{pi}) &= \sigma_y(0) + \int H' d\bar{\varepsilon}^{pi} \\
&= \sigma_y(0) + m \int H' d\bar{\varepsilon}^p
\end{aligned}
\tag{7.66}
$$

Similarly, the evolution of the back stress (cf. Eqs. 7.60–7.62) is

$$
\begin{aligned}
\dot{\rho} &= \dot{\mu}(\bar{\varepsilon}^{pk})(\sigma - \rho) \\
&= \frac{\bar{H}}{\sigma_y}(1-m)\dot{\varepsilon}^p(\sigma - \rho)
\end{aligned}
\tag{7.67}
$$

Finally, the parameter ψ can be obtained (see details in Appendix D.3) from the equation

$$
\psi = \frac{\partial \bar{f}}{\partial \sigma} : \mathbf{D}^e : \frac{\partial \bar{f}}{\partial \sigma} + H^* h_{\varepsilon,\lambda}
\tag{7.68}
$$

where the parameter H^* is a weighted average of the isotropic and kinematic hardening parameters, given by

$$
H^* = mH' + (1-m)\bar{H}
\tag{7.69}
$$

and the term $h_{\varepsilon,\lambda}$ is determined from Eq. (7.50). Clearly, using this approach, we see that the isotropic and kinematic hardening formulations are special cases of the mixed hardening formulation. In practice, this unified formulation facilitates the writing of general-purpose computer code.

7.8 Closure

Elasto–plastic flow theory is a difficult subject, as perhaps evidenced by this lengthy description of the bare essentials. This exposition has necessarily skipped over much of the detail, sacrificing rigor and implicitly invoking certain assumptions without discussion of their merit. Our primary aim, however, has been to derive the elasto–plastic incremental stress–strain relationships for some common yield functions and for reasonably realistic hardening rules. Although we have left the final results in matrix form, the last step (to evaluate the matrix coefficients term by term) should now be within easy reach. Of course, we have already undertaken this final step and the results can be extracted from the computer code listing in Chapter Ten.

CHAPTER EIGHT

Boundary Integral Equations in Elasto–Plasticity

8.1 Introduction

In this chapter, we describe a boundary element formulation for elasto–plastic deformation problems for materials governed by the constitutive equations described earlier. In this formulation, we discretize the yield region into interior cells to evaluate the initial stresses. For many problems, the yield region is fairly localized, and so the boundary element method is particularly efficient in these cases.

8.2 Boundary Integral Equations

Formulations for elasto–plastic boundary element methods date back to the early 1970s. However, the first accurate descriptions may be credited to Mukherjee (1977) and Bui (1978), and Telles & Brebbia (1980) presented the expressions for internal stresses. Unlike the linear case, here we must deal with incremental quantities, that is, increments of stress and strain, denoted by the superposed period (for example, $\dot{\sigma}$).

We begin by considering two elastic equilibrium states in the domain Ω, with boundary Γ, characterized by $(\dot{\sigma}_{ij}^{e}, \dot{\varepsilon}_{ij})$ and $(\sigma_{ij}^{*}, \varepsilon_{ij}^{*})$, noting that the latter are not incremental quantities. Using Hooke's law, and multiplying both sides by ε_{ij}^{*}, we obtain

$$
\begin{aligned}
\dot{\sigma}_{ij}^{e}\varepsilon_{ij}^{*} &= \lambda\delta_{ij}\dot{\varepsilon}_{kk}\varepsilon_{ij}^{*} + 2G\dot{\varepsilon}_{ij}\varepsilon_{ij}^{*} \\
&= \lambda\dot{\varepsilon}_{kk}\varepsilon_{mm}^{*} + 2G\dot{\varepsilon}_{ij}\varepsilon_{ij}^{*} \\
&= (\lambda\delta_{ij}\varepsilon_{mm}^{*} + 2G\varepsilon_{ij}^{*})\dot{\varepsilon}_{ij} \\
&= \sigma_{ij}^{*}\dot{\varepsilon}_{ij}
\end{aligned}
\tag{8.1}
$$

From this reciprocal identity, the following integral statement results:

$$
\int_{\Omega}\dot{\sigma}_{ij}^{e}\varepsilon_{ij}^{*}d\Omega = \int_{\Omega}\sigma_{ij}^{*}\dot{\varepsilon}_{ij}d\Omega
\tag{8.2}
$$

Recalling the decomposition of the strain increments into elastic and plastic parts (see Eq. 7.30), we now write the stress increment in the form

$$\dot{\sigma}_{ij} = \dot{\sigma}_{ij}^e - \dot{\sigma}_{ij}^P \tag{8.3}$$

where $\dot{\sigma}_{ij}^e$ is a notional elastic stress increment (corresponding to the strain increment) and $\dot{\sigma}_{ij}^P$ is termed the "initial" stress increment. However, the negative sign in this equation emphasizes the difference between this rather artificial decomposition of stress increments and that for strains. Since, by definition,

$$\dot{\sigma}_{ij}^e = D_{ijkl}^e \dot{\varepsilon}_{kl} \tag{8.4}$$

it follows from Eq. (7.31) that the initial stress increment is related to the plastic strain increment by

$$\dot{\sigma}_{ij}^P = D_{ijkl}^e \dot{\varepsilon}_{kl}^P \tag{8.5}$$

Thus, the initial stress increment is related, via the *elastic* constitutive relationship, to the irrecoverable (plastic) component of the strain increment.

We now recall the equilibrium equations

$$\dot{\sigma}_{ij,j} + b_i = 0$$
$$\sigma_{ij,j}^* + b_i^* = 0 \tag{8.6}$$

the traction–stress relationships

$$\dot{t}_i = \dot{\sigma}_{ij} n_j$$
$$t_i^* = \sigma_{ij}^* n_j \tag{8.7}$$

and the strain–displacement relationship

$$\dot{\varepsilon}_{ij} = (\dot{u}_{i,j} + \dot{u}_{j,i})/2$$
$$\varepsilon_{ij}^* = (u_{i,j}^* + u_{j,i}^*)/2 \tag{8.8}$$

Now substituting these equations into Eq. (8.2), and integrating by parts, we obtain

$$\int_{\Omega} \dot{\sigma}_{ij}^e \varepsilon_{ij}^* d\Omega = \int_{\Omega} \dot{\sigma}_{ij} u_{i,j}^* d\Omega + \int_{\Omega} \dot{\sigma}_{ij}^P \varepsilon_{ij}^* d\Omega$$
$$= \int_{\Gamma} u_i^* \dot{t}_i d\Gamma + \int_{\Omega} u_i^* b_i d\Omega + \int_{\Omega} \varepsilon_{ij}^* \dot{\sigma}_{ij}^P d\Omega \tag{8.9}$$

Similarly, the right-hand side of Eq. (8.2) can be integrated to obtain

$$\int_{\Omega} \sigma_{ij}^* \dot{\varepsilon}_{ij} d\Omega = \int_{\Gamma} t_i^* \dot{u}_i d\Gamma + \int_{\Omega} b_i^* \dot{u}_i d\Omega \tag{8.10}$$

Taken together, these equations yield

$$\int_{\Gamma} u_i^* \dot{t}_i d\Gamma + \int_{\Omega} u_i^* b_i d\Omega + \int_{\Omega} \varepsilon_{ij}^* \dot{\sigma}_{ij}^P d\Omega = \int_{\Gamma} t_i^* \dot{u}_i d\Gamma + \int_{\Omega} b_i^* \dot{u}_i d\Omega \tag{8.11}$$

Now, in a similar fashion to the approach followed in Chapter Three, we take the asterisked quantities to be those generated by unit loads in an infinite elastic solid, that is, Kelvin's fundamental solutions U_{ij} and T_{ij}. It follows that

$$\dot{u}_i(p) = \int_\Gamma U_{ij}(p, Q)\dot{t}_j(Q)d\Gamma(Q) - \int_\Gamma T_{ij}(p, Q)\dot{u}_j(Q)d\Gamma(Q)$$

$$+ \int_\Omega U_{ij}(p, q)b_j(q)d\Omega(q) + \int_\Omega E_{ijk}(p, q)\dot{\sigma}^P_{jk}(q)d\Omega(q)$$

$$(8.12)$$

where p and q denote the source point and field point, respectively, for internal points, and P and Q are the corresponding points on the boundary. The strain distribution E_{ijk} corresponding to the displacement field U_{ij} is obtained from the equation

$$E_{ijk}(p, q) = (U_{ij,k} + U_{ik,j})/2$$

$$= \frac{-1}{8\alpha\pi(1 - \nu)Gr^\alpha}\{(1 - 2\nu)(r_{,k}\delta_{ij} + r_{,j}\delta_{ik}) - r_{,i}\,\delta_{jk} + \beta r_{,i}\,r_{,j}\,r_{,k}\}$$

$$(8.13)$$

where the constants $\alpha = 2$ and $\beta = 3$ in three dimensions (but $\alpha = 1$ and $\beta = 2$ in plane strain). Equation (8.12) is only applicable for internal points. For points on the boundary, the limiting form is obtained by allowing the source point to approach the boundary (as described in Chapter Three), which yields

$$c_{ij}(P)\dot{u}_j(P) + \int_\Gamma T_{ij}(P, Q)\dot{u}_j(Q)d\Gamma(Q) = \int_\Gamma U_{ij}(P, Q)\dot{t}_j(Q)d\Gamma(Q)$$

$$+ \int_\Omega U_{ij}(P, q)b_j(q)d\Omega(q)$$

$$+ \int_\Omega E_{ijk}(P, q)\dot{\sigma}^P_{jk}(q)d\Omega(q)$$

$$(8.14)$$

where $c_{ij}(P)$ is related to the geometry at the boundary. For a smooth boundary (at P), $c_{ij}(P) = \delta_{ij}/2$. Apart from the additional domain integral, involving initial stresses, this boundary integral equation is the same as that obtained earlier for linear elastic analysis. To solve this equation, we have to determine the initial stresses and this requires an additional integral equation for the stresses within the domain. In passing, it is worth noting that an alternative formulation is possible in which "initial strains" are employed rather than initial stresses. Details are given in Appendix E.

8.3 Internal Stress Integral Equations

To determine the stresses within the domain, we begin with the displacements, as expressed by Eq. (8.12). Differentiation of these displacements yields the strains and then, using Hooke's law, the stresses are recovered. Some care is needed to accomplish this correctly and the main steps are presented here for completeness.

From Hooke's law and the strain–displacement relationships, we have

$$\dot{\sigma}_{ij}^e = \frac{2Gv}{1-2v}\dot{u}_{k,k}\delta_{ij} + G(\dot{u}_{i,j} + \dot{u}_{j,i}) \tag{8.15}$$

The displacement derivatives in this equation must now be obtained from Eq. (8.12). Since we are calculating stresses at p, the differentiation must be carried out with respect to the coordinates at p, a point that is highlighted in the following equations by the use of the superscripted variable x^p. Within the domain, the spatial derivatives of displacements are

$$\frac{\partial \dot{u}_i(p)}{\partial x_m^p} = \int_\Gamma \frac{\partial U_{ij}(p,Q)}{\partial x_m^p}\dot{t}_j(Q)d\Gamma(Q) - \int_\Gamma \frac{\partial T_{ij}(p,Q)}{\partial x_m^p}\dot{u}_j(Q)d\Gamma(Q)$$

$$+ \int_\Omega \frac{\partial U_{ij}(p,q)}{\partial x_m^p}b_j(q)d\Omega(q) + \frac{\partial}{\partial x_m^p}\int_\Omega E_{ijk}(p,q)\dot{\sigma}_{jk}^p(q)d\Omega(q) \tag{8.16}$$

All of these differentials, except the last, can be evaluated without difficulty. However, the last domain integral is strongly singular and a "jump" term arises during this process. To deal with this integral, a small sphere Ω_ε with radius ε, and centered at p, is separated from Ω, as depicted in Fig. 8.1. This term is separated into two parts:

$$\frac{\partial}{\partial x_m^p}\int_\Omega E_{ijk}(p,q)\dot{\sigma}_{jk}^p(q)d\Omega(q) = \lim_{\varepsilon \to 0}\int_{\Omega-\Omega_\varepsilon} \frac{\partial E_{ijk}(p,q)}{\partial x_m^p}\dot{\sigma}_{jk}^p(q)d\Omega(q)$$

$$+ \dot{\sigma}_{jk}^p(p)\lim_{\varepsilon \to 0}\int_{\Omega_\varepsilon} \frac{\partial E_{ijk}(p,q)}{\partial x_m^p}d\Omega(q) \tag{8.17}$$

in which $\lim_{\varepsilon \to 0}\dot{\sigma}_{jk}^p(q) = \dot{\sigma}_{jk}^p(p)$ has been used. Noting that $\partial(\cdot)/\partial x_m^p = -\partial(\cdot)/\partial x_m^q$ and using Gauss's theorem, we can integrate the last integral analytically to get

$$\int_{\Omega_\varepsilon} \frac{\partial E_{ijk}(p,q)}{\partial x_m^p}d\Omega(q) = -\int_\varepsilon E_{ijk}(p,Q)n_m(Q)d\Gamma(Q) \tag{8.18}$$

where n is the unit outward normal to the sphere of exclusion. For the

Figure 8.1: A small sphere Ω_ε cut out from Ω.

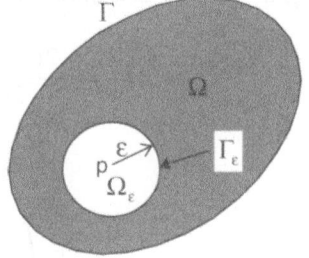

three-dimensional case, we can show (Appendix E) that

$$\int_{\Omega_\varepsilon} \frac{\partial E_{ijk}(p,q)}{\partial x_m^p} d\Omega(q) = \frac{-1}{30(1-\nu)G}\{\delta_{im}\delta_{jk} - (4-5\nu)(\delta_{ij}\delta_{km} + \delta_{ik}\delta_{jm})\}$$

(8.19)

and for the plane strain case, the corresponding result is

$$\int_{\Omega_\varepsilon} \frac{\partial E_{ijk}(p,q)}{\partial x_m^p} d\Omega(q) = \frac{-1}{16(1-\nu)G}\{\delta_{im}\delta_{jk} - (3-4\nu)(\delta_{ij}\delta_{km} + \delta_{ik}\delta_{jm})\}$$

(8.20)

Combining these results with the differentiations indicated in Eq. (8.16), we can express Eq. (8.15) in the form

$$\dot{\sigma}_{ij}(p) = \int_\Gamma U_{ijk}(p,Q)\dot{t}_k(Q)d\Gamma(Q) - \int_\Gamma T_{ijk}(p,Q)\dot{u}_k(Q)d\Gamma(Q)$$

$$+ \int_\Omega U_{ijk}(p,q)\dot{b}_k(q)d\Omega(q) + \int_\Omega E_{ijkl}(p,q)\dot{\sigma}_{kl}^p(q)d\Omega(q) \quad (8.21)$$

$$+ F_{ijkl}^\sigma \dot{\sigma}_{kl}^p(p)$$

where

$$U_{ijk} = \frac{1}{4\alpha\pi(1-\nu)r^\alpha}\{(1-2\nu)(r_{,i}\,\delta_{jk} + r_{,j}\,\delta_{ki} - r_{,k}\,\delta_{ij}) + \beta r_{,i}\,r_{,j}\,r_{,k}\}$$

(8.22)

$$T_{ijk} = \frac{G}{2\alpha\pi(1-\nu)r^\beta}\{\beta r_{,m}\,n_m[(1-2\nu)r_{,k}\,\delta_{ij}$$

$$+ \nu(r_{,j}\,\delta_{ki} + r_{,i}\,\delta_{jk}) - \gamma r_{,i}\,r_{,j}\,r_{,k}]$$

$$- (1-4\nu)n_k\delta_{ij} + \beta\nu(n_i r_{,j}\,r_{,k} + n_j r_{,i}\,r_{,k})$$

$$+ (1-2\nu)(\beta n_k r_{,i}\,r_{,j} + n_j\delta_{ik} + n_i\delta_{jk})\} \quad (8.23)$$

and

$$E_{ijkl} = \frac{1}{4\alpha\pi(1-\nu)r^\beta}\{(1-2\nu)(\delta_{ik}\delta_{lj} + \delta_{jk}\delta_{li} - \delta_{ij}\delta_{kl} + \beta\delta_{ij}r_{,k}\,r_{,l})$$

$$+ \beta\nu(\delta_{li}r_{,j}\,r_{,k} + \delta_{jk}r_{,l}\,r_{,i} + \delta_{ik}r_{,l}\,r_{,j} + \delta_{jl}r_{,i}\,r_{,k})$$

$$+ \beta\delta_{kl}r_{,i}\,r_{,j} - \beta\gamma r_{,i}\,r_{,j}\,r_{,k}r_{,l}\} \quad (8.24)$$

and where the constants $\alpha = 2$, $\beta = 3$, and $\gamma = 5$ in three dimensions (but $\alpha = 1$, $\beta = 2$, and $\gamma = 4$ under plane strain conditions). The free terms are

$$F_{ijkl}^\sigma \dot{\sigma}_{kl}^p(p) = \frac{2G\nu}{1-2\nu}\delta_{ij}\dot{\sigma}_{kl}^p(p)\int_{\Omega_\varepsilon} \frac{\partial E_{mkl}(p,q)}{\partial x_m^p} d\Omega(q)$$

$$+ G\dot{\sigma}_{kl}^p(p)\left[\int_{\Omega_\varepsilon} \frac{\partial E_{ikl}(p,q)}{\partial x_j^p} d\Omega(q)\right.$$

$$+ \left.\int_{\Omega_\varepsilon} \frac{\partial E_{jkl}(p,q)}{\partial x_i^p} d\Omega(q)\right] - \dot{\sigma}_{ij}^p(p)$$

(8.25)

Thus, from Eqs. (8.19) and (8.20), we obtain, for three-dimensional conditions,

$$F_{ijkl}^{\sigma} = \frac{-1}{30(1-v)}\{(7-5v)(\delta_{ik}\delta_{jl}+\delta_{il}\delta_{jk})+(2-10v)\delta_{ij}\delta_{kl}\} \tag{8.26}$$

and, for plane strain,

$$F_{ijkl}^{\sigma} = \frac{-1}{8(1-v)}\{(\delta_{ik}\delta_{jl}+\delta_{il}\delta_{jk})+(1-4v)\delta_{ij}\delta_{kl}\} \tag{8.27}$$

Initial strain equivalents to these equations are given in Appendix E. As usual, the singular domain integrals are interpreted in the Cauchy principal value sense. Those involving the initial stresses are strongly singular and require special treatment to integrate them. A number of methods can be employed for this purpose (e.g., Dallner & Kuhn, 1993), but here a particularly efficient technique developed by the authors (Gao & Davies, 2000b) is employed.

8.4 Integration of Strongly Singular Domain Integrals

The essence of the technique is the isolation of the singularity and its transformation into a (local) boundary integral. The singularity that arises in the strongly singular domain integral can be isolated by rewriting the integral in the form

$$\int_{\Omega} E_{ijkl}(p,q)\dot{\sigma}_{kl}^{P}(q)d\Omega(q) = \int_{\Omega} E_{ijkl}(p,q)[\dot{\sigma}_{kl}^{P}(q)-\dot{\sigma}_{kl}^{P}(p)]d\Omega(q)$$

$$+\dot{\sigma}_{kl}^{P}(p)\left[\int_{\Omega} E_{ijkl}(p,q)d\Omega(q)\right] \tag{8.28}$$

The first integral on the right-hand side is weakly singular and can be integrated numerically. The strong singularity has been transferred to the last integral and can be dealt with semianalytically, as follows. First, for convenience, we rewrite the kernel E_{ijkl} in the form

$$E_{ijkl} = \frac{\Psi_{ijkl}}{r^{\beta}} \tag{8.29}$$

where $\beta = 3$ in three dimensions ($\beta = 2$ in plane strain and plane stress) and

$$\Psi_{ijkl} = \frac{1}{4\alpha\pi(1-v)}\{(1-2v)(\delta_{ik}\delta_{lj}+\delta_{jk}\delta_{li}-\delta_{ij}\delta_{kl}+\beta\delta_{ij}r_{,k}r_{,l})$$

$$+\beta v(\delta_{li}r_{,j}r_{,k}+\delta_{jk}r_{,l}r_{,i}+\delta_{ik}r_{,l}r_{,j}+\delta_{jl}r_{,i}r_{,k})$$

$$+\beta\delta_{kl}r_{,i}r_{,j}-\beta\gamma r_{,i}r_{,j}r_{,k}r_{,l}\} \tag{8.30}$$

Next, we divide the region Ω into two parts, Ω_s and Ω_r, as depicted in Fig. 8.2. The region Ω_s, bounded by Γ_s and including p, we refer to as the singular domain. In practice, the boundary Γ_s is formed by the outer boundaries of the cells surrounding p. The external (regular) region Ω_r is nonsingular. Since the domain integral is interpreted in the Cauchy principal value sense, we cut out a small sphere, with radius ε, from the region Ω_s, as depicted in Fig. 8.3. Using a local spherical

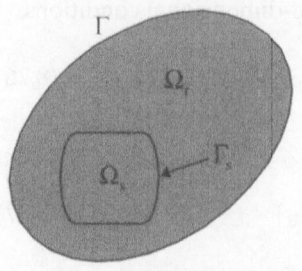

Figure 8.2: Division of domain into regular and singular regions.

coordinate system (Appendix E) with origin at p, we can write the domain integration over the sphere Ω_s as

$$\int_{\Omega_s} E_{ijkl}d\Omega = \int_0^\pi \int_0^{2\pi} \left[\lim_{\varepsilon \to 0} \int_\varepsilon^{r(\Gamma_s)} \frac{1}{r}dr \right] \Psi_{ijkl} \sin\theta d\phi d\theta$$

$$= \int_0^\pi \int_0^{2\pi} \log_e r(\Gamma_s)\Psi_{ijkl} \sin\theta d\phi d\theta$$

$$- \lim_{\varepsilon \to 0} \log_e \varepsilon \int_0^\pi \int_0^{2\pi} \Psi_{ijkl} \sin\theta d\phi d\theta \tag{8.31}$$

Using results obtained in Appendix E (namely, Eqs. E.7–E.9), we can prove that

$$\int_0^\pi \int_0^{2\pi} \Psi_{ijkl} \sin\theta d\phi d\theta = 0 \tag{8.32}$$

Further, referring to Fig. 8.4, we have

$$dS = d\Gamma \cos\varphi = d\Gamma \frac{r_i n_i}{r} \tag{8.33}$$

where φ is the angle between the normal of the differential element of the spherical surface dS directed along the r direction and the cell boundary surface $d\Gamma$ with normal n. Hence, we obtain

$$\sin\theta d\phi d\theta = \frac{r_i n_i}{r^3}d\Gamma \tag{8.34}$$

Taking these equations together, we obtain, finally,

$$\int_{\Omega_s} E_{ijkl}d\Omega = \int_{\Gamma_s} E_{ijkl}r_m n_m \log_e r d\Gamma \tag{8.35}$$

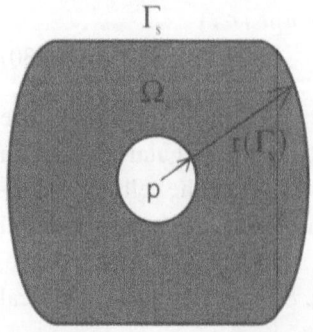

Figure 8.3: Sphere of exclusion (of radius ε) within Ω_s.

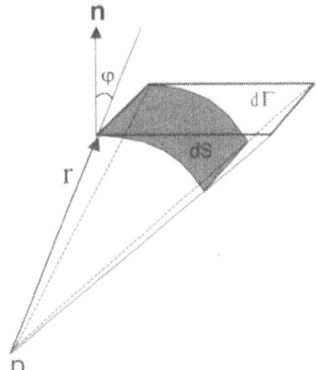

Figure 8.4: Relationship between spherical surface (dS)
and cell boundary ($d\Gamma$).

An analogous result can be obtained using the initial strain approach (Appendix E,
Section E.5).

The significance of Eq. (8.35) is that we have transformed the strongly singular
domain integral into a nonsingular boundary integral; in a numerical implementation, the boundary would be the outer surfaces of the cells surrounding the
source point. This integral is readily calculated numerically using standard Gauss
quadrature. For two-dimensional problems (Gao & Davies, 2000b) precisely the
same result is obtained (as briefly described in Appendix E, Section E.4), which
facilitates the development of general-purpose computer code.

8.5 Closure

In this chapter, we have summarized the principal equations that form the basis
of the boundary element algorithm and sketched in some of the details of their
derivations. These equations share much in common with those derived earlier
for linear analyses, but here the interior field quantities (initial strains or initial
stresses) are unknown a priori. This complication introduces considerable difficulties, and the development of efficient robust code to close the system equations is
no simple matter. Another difficulty in the nonlinear formulation is that strongly
singular volume integrals arise, but the novel semianalytical integration technique
advocated here offers the promise of resolving this problem. How these theoretical formulations are translated into practical numerical techniques is the subject
of Chapter Nine.

CHAPTER NINE

Numerical Implementation

9.1 Introduction

In this chapter, we describe the numerical solution of the elasto–plastic boundary integral equations. One obvious difference between the linear and nonlinear algorithms is that the discretization now requires division of the yield region into cells. Ultimately, a set of algebraic equations is obtained in which the plastic multipliers are the primary unknowns. These equations are solved using an incremental Newton–Raphson iterative scheme.

9.2 Domain Discretization

In this nonlinear analysis, the boundary integral equations are augmented by domain integrals, which necessitate domain discretization. However, only the yield region needs to be discretized into cells because the domain integrals are zero elsewhere. The domain discretization is illustrated by the two-dimensional example depicted in Fig. 9.1.

For this and other reasons, the domain integration is quite different in character from that employed in the finite element method. One complication is that the extent of the yield region is not known a priori and consequently it is usual to assume generous proportions for it in pilot studies. In other words, the internal boundary between the cells and the exterior nondiscretized region must lie wholly within the elastic region.

We assume that the yield region is divided into N_c iso-parametric cells: in each cell, the geometry and initial stresses are interpolated between the nodes using shape functions, that is,

$$x_i = \sum_{\alpha=1}^{M} N_\alpha(\xi_k) x_i^\alpha \tag{9.1}$$

$$\dot{\sigma}_{ij}^p = \sum_{\alpha=1}^{M} N_\alpha(\xi_k) \dot{\sigma}_{ij}^{p\alpha} \tag{9.2}$$

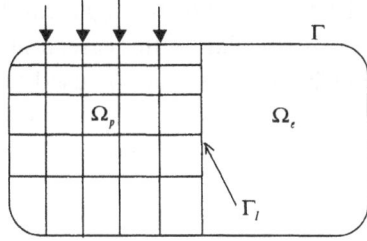

Figure 9.1: Domain discretization in nonlinear problems.

where M is the number of cell nodes, $\dot{\sigma}_{ij}^{p\alpha}$ is the ijth component of the initial stress at node α, and $N_\alpha(\xi_k)$ are the shape functions (Appendix B). Thus, the domain integrals in Eqs. (8.14) and (8.21) can be written as

$$\int_\Omega E_{ijk}(p,q)\dot{\sigma}_{jk}^p(q)d\Omega(q) = \sum_{c=1}^{N_c}\sum_{\alpha=1}^{M} E_{ijk}^{c\alpha}\dot{\sigma}_{jk}^{p\alpha} \qquad (9.3a)$$

$$\int_\Omega E_{ijkl}(p,q)\dot{\sigma}_{kl}^p(q)d\Omega(q) = \sum_{c=1}^{N_c}\sum_{\alpha=1}^{M} E_{ijkl}^{c\alpha}\dot{\sigma}_{kl}^{p\alpha} \qquad (9.3b)$$

where

$$E_{ijk}^{c\alpha} = \int_{\Omega_c} E_{ijk}(p,q)N_\alpha(q)d\Omega(q) \qquad (9.4a)$$

$$E_{ijkl}^{c\alpha} = \int_{\Omega_c} E_{ijkl}(p,q)N_\alpha(q)d\Omega(q) \qquad (9.4b)$$

in which Ω_c is the domain of the cell c. When the source point p is not one of the cell's nodes, Gauss quadrature can be used to evaluate the integrals in Eq. (9.4). In three dimensions, we obtain

$$E_{ijk}^{c\alpha} = \int_{-1}^{1}\int_{-1}^{1}\int_{-1}^{1} E_{ijk}(p,q)N_\alpha(\xi_1,\xi_2,\xi_3)J_c(\xi_1,\xi_2,\xi_3)d\xi_1 d\xi_2 d\xi_3 \qquad (9.5a)$$

$$E_{ijkl}^{c\alpha} = \int_{-1}^{1}\int_{-1}^{1}\int_{-1}^{1} E_{ijkl}(p,q)N_\alpha(\xi_1,\xi_2,\xi_3)J_c(\xi_1,\xi_2,\xi_3)d\xi_1 d\xi_2 d\xi_3 \qquad (9.5b)$$

where J_c is the Jacobian of the transformation,

$$\begin{aligned} J_c(\xi_1,\xi_2,\xi_3) &= \left|\frac{\partial(x,y,z)}{\partial(\xi_1,\xi_2,\xi_3)}\right| \\ &= \begin{vmatrix} \dfrac{\partial x}{\partial\xi_1} & \dfrac{\partial y}{\partial\xi_1} & \dfrac{\partial z}{\partial\xi_1} \\[2mm] \dfrac{\partial x}{\partial\xi_2} & \dfrac{\partial y}{\partial\xi_2} & \dfrac{\partial z}{\partial\xi_2} \\[2mm] \dfrac{\partial x}{\partial\xi_3} & \dfrac{\partial y}{\partial\xi_3} & \dfrac{\partial z}{\partial\xi_3} \end{vmatrix} \end{aligned} \qquad (9.6)$$

Of course, in two dimensions (quadrilateral cells), the triple integrals reduce to double integrals and the Jacobian matrix is of rank 2. If the dimensions of adjacent cells are significantly different, the integrals in Eqs. (9.5) become nearly singular. In this case, a subdivision technique (analogous to that described in Chapter 4) should be used to preserve accuracy. For singular cells, the weak singularities in Eq. (9.5a) and the strong singularities in Eq. (9.5b) require special treatment, as described in the following two sections.

9.3 Weakly Singular Domain Integrals

Weak singularities can be dealt with quite easily by, in effect, using numerical quadrature schemes tailored for such singular integrals. This can be done by distorting standard quadrature schemes, through appropriate coordinate transformations. The principle is illustrated, for a two-dimensional cell, in Fig. 9.2a, which is distorted into a triangle by merging all the nodes on a side into one. The resulting "degenerate" cell has the useful property that the Jacobian of the mapping from the original cell to its degenerate form is proportional to r, where r is the distance from the common node, as demonstrated in Appendix C. Thus, if the singularity is centered on this common node, it is nullified by the coordinate transformation of the quadrature scheme. Similarly, for a three-dimensional cell (Fig. 9.2b), if all nodes on a surface are merged into one, then the Jacobian of the transformation becomes proportional to r^2.

Based on these observations, we can eliminate weak singularities by subdividing cells into subcells and then carrying out subcell mappings into intrinsic subcells. The subdivision into subcells, centered on the singularity, is depicted in Fig. 9.3. Here, we deal explicitly only with the three-dimensional case; the two-dimensional case follows naturally. For a corner node, three such subcells are created, but four are needed for mid-side nodes.

Each subcell, centered on the source point p, is then mapped into a unit cube, as depicted in Fig. 9.4. Using this subdivision technique, Eq. (9.5a) can be expressed as

$$
E_{ijk}^{c\alpha} = \sum_{s}^{N_c^s} \int_{-1}^{1} \int_{-1}^{1} \int_{-1}^{1} E_{ijk}(p,q) N_\alpha(\xi_1, \xi_2, \xi_3)
$$
$$
\times J_c(\xi_1, \xi_2, \xi_3) J_c^s(\xi_1', \xi_2', \xi_3') d\xi_1' d\xi_2' d\xi_3' \tag{9.7}
$$

where N_c^s is the number of the subcells and J_c^s is the Jacobian of transformation from the original local coordinates to the new subcell intrinsic coordinate system,

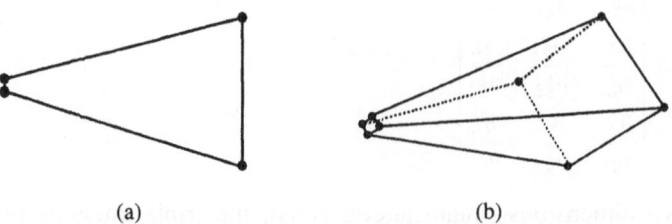

(a)   (b)

Figure 9.2: Degenerate elements in (a) two dimensions and (b) three dimensions.

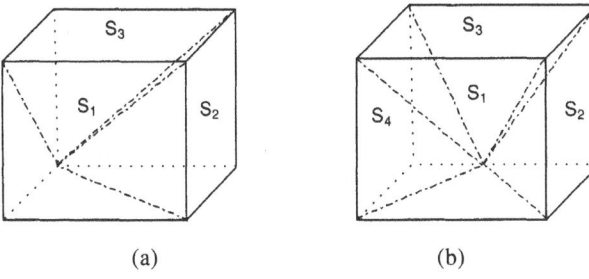

Figure 9.3: Cell subdivision: (a) p at corner node; (b) p at mid-side node.

that is,

$$J_c^s(\xi_1', \xi_2', \xi_3') = \left| \frac{\partial(\xi_1, \xi_2, \xi_3)}{\partial(\xi_1', \xi_2', \xi_3')} \right| \tag{9.8}$$

in which the original local coordinates are interpolated from the eight-noded linear cell. Thus,

$$\xi_i = \sum_{\alpha=1}^{8} N_\alpha(\xi_1', \xi_2', \xi_3')\xi_i^\alpha \tag{9.9}$$

where N_α, $\alpha = 1, \ldots, 8$, are the shape functions (Appendix B) and ξ_i^α is the ith intrinsic coordinate at node α. Linear interpolation is sufficient here because the mapping from the intrinsic coordinates of the original cells to the intrinsic coordinates of the degenerate cells is linear. For the two-dimensional case (described in Appendix C), the Jacobian J_c^s tends to zero (here $O(r)$ as $r \to 0$), nullifying the weak singularity of $O(1/r)$ in the kernel E_{ijk} and permitting its accurate evaluation.

9.4 Strongly Singular Domain Integrals

The technique described above is not sufficient to eliminate the strong singularity of order $O(1/r^3)$ in the kernel E_{ijkl}. Consequently, for this kernel, we resort to the singularity isolation method described in the last chapter. Using this method,

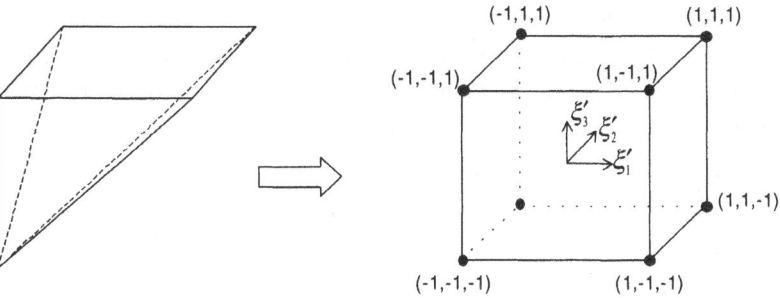

Figure 9.4: Geometrical mapping of a subcell onto a unit cube.

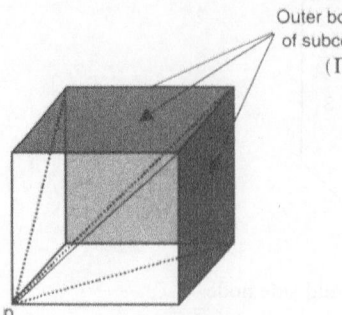

Outer boundaries
of subcells
(Γ_s)

Figure 9.5: Cell discretization for strongly singular integral.

p

we can decompose the singular integral (in Eq. 9.5b) as follows:

$$E_{ijkl}^{c\alpha} = \sum_{s=1}^{N_c^s} \int_{-1}^{1} \int_{-1}^{1} \int_{-1}^{1} E_{ijkl}(p,q)(N_\alpha - \delta_{\alpha p})J_c(\xi_1,\xi_2,\xi_3)J_c^s(\xi_1',\xi_2',\xi_3')d\xi_1'd\xi_2'd\xi_3'$$

$$+ \delta_{\alpha p} \sum_{\Gamma_s=1}^{N_c^s} \int_{-1}^{1} \int_{-1}^{1} E_{ijkl}(p,q)r_m(p,q)n_m(q)\log_e r(p,q)J_{\Gamma_s}(\xi_1,\xi_2)d\xi_1 d\xi_2$$

$$\text{(9.10)}$$

where N_c^s is the number of degenerate subcells, J_Γ^s is the Jacobian of the transformation over the cell boundaries, and

$$\delta_{\alpha p} = \begin{cases} 1 & \text{when } \alpha = p \\ 0 & \text{when } \alpha \neq p \end{cases} \tag{9.11}$$

The first integral is performed over the degenerate subcells, while the second is
carried out over the outer boundaries of these subcells. Figure 9.5 illustrates these
subcells and the corresponding outer boundaries for the case where the singularity
is at a corner node. The subcell discretization for a mid-side node follows naturally.

9.5 Boundary Stresses – Traction-Recovery Method

The stress kernel (Eq. 8.21) is hypersingular at the boundary and consequently
normal integration techniques are ineffective. We overcome this problem by
adopting a "traction-recovery" technique like that employed in Chapter Four, but
which now includes the initial stresses. First, we consider the three-dimensional
case and define a local Cartesian coordinate system x_i' as depicted in Fig. 9.6. The
"elastic" stress increments $\dot{\sigma}_{ij}^{\prime e}$ can be expressed, following the approach presented
in Chapter Four, in the form

$$\dot{\sigma}_{11}^{\prime e} = \frac{2G}{1-\nu}(\dot{\varepsilon}_{11}' + \nu\dot{\varepsilon}_{22}') + \frac{\nu}{1-\nu}\dot{\sigma}_{33}^{\prime e}$$

$$\dot{\sigma}_{22}^{\prime e} = \frac{2G}{1-\nu}(\dot{\varepsilon}_{22}' + \nu\dot{\varepsilon}_{11}') + \frac{\nu}{1-\nu}\dot{\sigma}_{33}^{\prime e} \tag{9.12}$$

$$\dot{\sigma}_{12}^{\prime e} = 2G\dot{\varepsilon}_{12}'$$

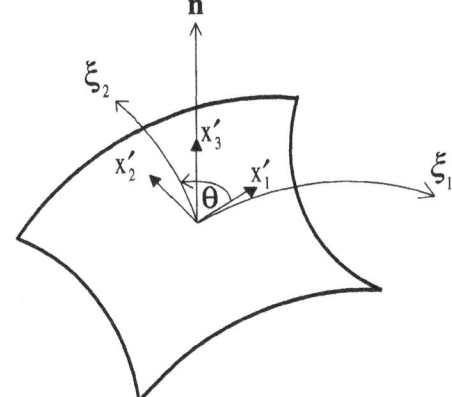

Figure 9.6: Local orthogonal set of axes over a boundary element.

where $\dot{\varepsilon}'_{11}$, $\dot{\varepsilon}'_{22}$, and $\dot{\varepsilon}'_{12}$ are determined using Eq. (4.69) and, by definition,

$$
\begin{aligned}
\dot{\sigma}'^e_{11} &= \dot{\sigma}'_{11} + \dot{\sigma}'^p_{11} \\
\dot{\sigma}'^e_{22} &= \dot{\sigma}'_{22} + \dot{\sigma}'^p_{22} \\
\dot{\sigma}'^e_{33} &= \dot{\sigma}'_{33} + \dot{\sigma}'^p_{33} \\
\dot{\sigma}'^e_{12} &= \dot{\sigma}'_{12} + \dot{\sigma}'^p_{12}
\end{aligned}
\tag{9.13}
$$

The transformation, from global to local axes, of the initial stresses $\dot{\sigma}'^p_{ij}$ can be expressed in the form

$$
\dot{\sigma}'^p_{ij} = L_{ik} L_{jl} \dot{\sigma}^p_{kl}
\tag{9.14}
$$

where L_{ij} are the direction cosines of the local coordinate system with respect to the global coordinate system (refer to Eqs. 4.76–4.77). After performing some lengthy manipulations, we obtain

$$
\begin{aligned}
\dot{\sigma}'_{11} &= \frac{2G}{1-\nu}\left(\frac{\partial \xi_K}{\partial x'_1} L_{1j} + \nu \frac{\partial \xi_K}{\partial x'_2} L_{2j}\right)\frac{\partial \dot{u}_j}{\partial \xi_K} + \frac{\nu}{1-\nu} L_{3j}\dot{t}_j \\
&\quad - \left(L_{1k}L_{1l} - \frac{\nu}{1-\nu} L_{3k}L_{3l}\right)\dot{\sigma}^p_{kl} \\
\dot{\sigma}'_{22} &= \frac{2G}{1-\nu}\left(\frac{\partial \xi_K}{\partial x'_2} L_{2j} + \nu \frac{\partial \xi_K}{\partial x'_1} L_{1j}\right)\frac{\partial \dot{u}_j}{\partial \xi_K} + \frac{\nu}{1-\nu} L_{3j}\dot{t}_j \\
&\quad - \left(L_{2k}L_{2l} - \frac{\nu}{1-\nu} L_{3k}L_{3l}\right)\dot{\sigma}^p_{kl} \\
\dot{\sigma}'_{12} &= G\left(\frac{\partial \xi_K}{\partial x'_1} L_{2j} + \frac{\partial \xi_K}{\partial x'_2} L_{1j}\right)\frac{\partial \dot{u}_j}{\partial \xi_K} - L_{1k}L_{2l}\dot{\sigma}^p_{kl}
\end{aligned}
\tag{9.15}
$$

Finally, using the transformation relation

$$
\dot{\sigma}_{mn} = L_{km} L_{1n} \dot{\sigma}'_{kl}
\tag{9.16}
$$

we obtain

$$
\dot{\sigma}_{mn} = A_{mnj\alpha}\dot{u}^\alpha_j + B_{mnj}\dot{t}_j + C_{mnkl}\dot{\sigma}^p_{kl}
\tag{9.17}
$$

where $A_{mnj\alpha}$ and B_{mnj} are identical to the coefficients obtained in Chapter Four, and

$$C_{mnkl} = -\sum_{i=1}^{3} L_{im}L_{in}L_{ik}L_{il} + \left(\frac{\nu}{1-\nu}\delta_{mn} + \frac{1-2\nu}{1-\nu}L_{3m}L_{3n}\right)L_{3k}L_{3l}$$
$$-\frac{1}{2}(L_{1m}L_{2n} + L_{2m}L_{1n})(L_{1k}L_{2l} + L_{2k}L_{1l}) \tag{9.18}$$

For the two-dimensional case, the results can be expressed in the same form as Eq. (9.17). The appropriate coefficients $A_{mnj\alpha}$ and B_{mnj} for plane strain and plane stress conditions may be found in Chapter Four; C_{mnkl} is

$$C_{mnkl} = -L_{1m}L_{1n}\left(\delta_{kl} - \frac{1}{1-\nu}L_{2k}L_{2l}\right) \tag{9.19}$$

where the subscripts range from 1 to 2, only. In addition, for plane strain problems, the following coefficient is also needed to calculate the stress in the third direction:

$$C_{33kl} = \frac{\nu}{1-\nu}L_{2k}L_{2l} - \delta_{3k}\delta_{3l} \tag{9.20}$$

Naturally, it may be preferred to encode the equations leading up to Eq. (9.17) rather than employ the explicit expressions obtained here for the coefficients $A_{mnj\alpha}$, B_{mnj}, and C_{mnkl}. Whichever method is adopted, the nodal stresses obtained are derived from the quantities defined over an element, and, in general, where tractions are continuous across a boundary node, these stresses may be averaged. However, for nodes where tractions are discontinuous, multiple nodes must be defined to resolve the discontinuities. In Fig 9.7, for example, the three different normal tractions (P_2, P_1, zero) at the central point can only be distinguished if three nodes (N_1, N_2, and N_3) are specified at this point. (We assume here that the tangential tractions follow a similar, or less demanding, pattern). The stresses at N_1 are averaged from the boundary elements E_1 and E_2, while those at nodes N_2 and N_3 are derived independently from elements E_3 and E_4, respectively.

Identification of multiple nodes at traction discontinuities is essential in the nonlinear analysis, even if the prescribed boundary conditions are the tractions themselves. In linear analysis, such boundary conditions can be dealt with, without recourse to multiple nodes, during assembly of the system equations. This is not

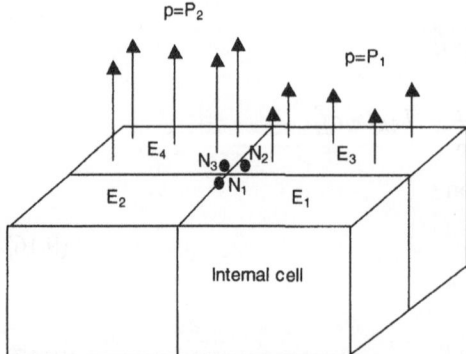

Figure 9.7: Multiple nodes at a traction discontinuity.

so in nonlinear analysis, where the stress discontinuities between cells must also be represented.

9.6 System Equations

Having discretized the boundary integral equations, we can now assemble the co-efficients into a set of system equations. However, because these equations contain terms (such as initial stresses) that are unknown a priori, the assembly needs to be done in a manner that facilitates isolation of these quantities.

9.6.1 Initial Stress Representation

Numerical integration of the domain and boundary integrals, for each boundary node in turn, yields the incremental system equations

$$[H]\{\dot{u}\} = [G]\{\dot{t}\} + [E^b]\{\dot{\sigma}^P\} \tag{9.21}$$

After substituting the boundary conditions into Eq. (9.21) and transferring all the boundary unknowns to the left-hand side and the known matrix products to the right-hand side, we obtain

$$[A^b]\{\dot{x}\} = \{\dot{y}^b\} + [E^b]\{\dot{\sigma}^P\} \tag{9.22}$$

where $\{\dot{x}\}$ are the boundary unknowns. Using the traction recovery method (Eq. 9.17), we can write the stresses at the boundary nodes in the matrix form

$$\{\dot{\sigma}^b\} = [H^b]\{\dot{u}\} + [G^b]\{\dot{t}\} + [\tilde{E}^b]\{\dot{\sigma}^P\} \tag{9.23a}$$

where $\{\dot{\sigma}^b\}$ are stresses at boundary nodes. The coefficient matrices $[H^b]$, $[G^b]$, and $[\tilde{E}^b]$ are extremely sparse. The stresses at internal nodes are obtained from the discretization of Eq. (8.21), which yields

$$\{\dot{\sigma}^I\} = [H^I]\{\dot{u}\} + [G^I]\{\dot{t}\} + [E^I]\{\dot{\sigma}^P\} \tag{9.23b}$$

where $\{\dot{\sigma}^I\}$ are the stresses at the internal nodes. Combining Eqs. (9.23a) and (9.23b) into one global equation and then substituting the prescribed boundary conditions leads to the equation

$$\{\dot{\sigma}\} = [A^\sigma]\{\dot{x}\} + \{\dot{y}^\sigma\} + [E^\sigma]\{\dot{\sigma}^P\} \tag{9.24}$$

where $\{\dot{\sigma}\}$ is a global stress vector, consisting of stresses at both boundary nodes and internal nodes. Although Eqs. (9.22) and (9.24) constitute a formal statement of the solution of nonlinear problems, they cannot be solved directly since the initial stress vector is unknown. The manner in which these two equations are solved differentiates between the various solution strategies. Before embarking on this crucial step, we note that once the system unknowns are calculated, internal displacements can be readily determined, if desired, from the discrete form of

Eq. (8.12), namely,

$$\{\dot{u}\} = [A^u]\{\dot{x}\} + \{\dot{y}^u\} + [E^u]\{\dot{\sigma}^p\} \tag{9.25}$$

Also, for those who prefer an "initial strain" approach, the analogous equations can be established via the stress–strain relationship (Eq. 8.5).

9.6.2 Plastic Multiplier Representation

The system equations described above (namely, Eqs. 9.22 and 9.24) can be solved directly by means of an incremental iterative scheme (Telles, 1983), but experience shows that this approach converges only slowly. A novel algorithm, based on the work of Banerjee et al. (1989), which casts the system equations in terms of plastic multipliers (Gao, 1999; Gao & Davies, 2000b), has proved to be both robust and computationally efficient. This method exploits the observation that the magnitudes of the plastic strains (a tensor quantity) are proportional to the plastic multiplier (a scalar quantity), and hence the number of unknowns may be substantially reduced by recasting the system equations in terms of this scalar quantity. In three dimensions, this approach reduces the number of system equations by a factor of 6.

We begin by recalling the definition of initial stress (Eq. 8.3), which we write here for a single node in the matrix form

$$\{\dot{\sigma}^e\} = \{\dot{\sigma}\} + \{\dot{\sigma}^p\} \tag{9.26}$$

Using the flow rule (Eq. 7.32) and the stress–strain relationship (Eq. 8.5), we obtain

$$\{\dot{\sigma}^p\} = \{d^f\}\dot{\lambda} \tag{9.27}$$

where

$$\{d^f\} = [D^e]\left\{\frac{\partial f}{\partial \sigma}\right\} \tag{9.28}$$

Now, from Eq. (7.40) and Eq. (8.4) we obtain

$$\dot{\lambda} = \{\nabla f_\psi\}^T\{\dot{\sigma}^e\} \tag{9.29}$$

where T denotes the vector transpose and

$$\{\nabla f_\psi\} = \frac{1}{\psi}\left\{\frac{\partial f}{\partial \sigma}\right\} \tag{9.30}$$

If we assume that N nodes have yielded, then a global initial stress vector can be formed from Eq. (9.27) in the form

$$\{\dot{\sigma}^p\} = [d^f]\{\dot{\lambda}\} \tag{9.31}$$

where $[d^f]$ is a $6N \times N$ diagonally dominant sparse matrix in three dimensions. For plane strain and plane stress problems, the number of rows will be $4N$ and

$3N$, respectively. Inverting Eq. (9.22) and using Eq. (9.31), we get

$$\{\dot{x}\} = \{\dot{y}^c\} + [A^c][d^f]\{\dot{\lambda}\} \tag{9.32}$$

where

$$\{\dot{y}^c\} = [A^b]^{-1}\{\dot{y}^b\} \tag{9.33}$$

$$[A^c] = [A^b]^{-1}[E^b] \tag{9.34}$$

Substituting Eqs. (9.31) and (9.32) into Eq. (9.24) leads to

$$\{\dot{\sigma}\} = \{\dot{y}^e\} + [E][d^f]\{\dot{\lambda}\} \tag{9.35}$$

where

$$\{\dot{y}^e\} = \{\dot{y}^\sigma\} + [A^\sigma]\{\dot{y}^c\} \tag{9.36}$$

and

$$[E] = [E^\sigma] + [A^\sigma][A^c] \tag{9.37}$$

Now, substituting Eqs. (9.35) and (9.31) into the global form of Eq. (9.26), and putting the result into Eq. (9.29), we obtain

$$\{\dot{\lambda}\} = [\nabla f_\psi](\{\dot{y}^e\} + [E][d^f]\{\dot{\lambda}\} + [d^f]\{\dot{\lambda}\}) \tag{9.38}$$

where $[\nabla f_\psi]$ is a $N \times 6N$ diagonally dominant matrix for three-dimensional problems, formed by assembling the nodal vectors $\{\nabla f_\psi\}^T$ from Eq. (9.30). Rearranging Eq. (9.38), we obtain the system equations

$$[A^\lambda]\{\dot{\lambda}\} = \{\dot{y}^f\} \tag{9.39}$$

where

$$[A^\lambda] = [I] - [\nabla f_\psi][C][d^f] \tag{9.40}$$

$$\{\dot{y}^f\} = [\nabla f_\psi]\{\dot{y}^e\} \tag{9.41}$$

and $[C]$ is a constant matrix

$$[C] = [I] + [E] \tag{9.42}$$

Once the plastic multipliers $\{\dot{\lambda}\}$ are obtained from Eq. (9.39), the increments of the boundary unknowns and stresses can be computed using Eqs. (9.32) and (9.35), respectively. And the interior displacements, if required, can be calculated, through substituting Eqs. (9.32) and (9.31) into Eq. (9.25), using the equation

$$\{\dot{u}\} = \{\dot{y}_u^e\} + [E_u^I][d^f]\{\dot{\lambda}\} \tag{9.43}$$

where

$$\{\dot{y}_u^e\} = \{\dot{y}^u\} + [A^u]\{\dot{y}^c\} \tag{9.44}$$

$$[E_u^I] = [E^u] + [A^u][A^c] \tag{9.45}$$

It is critically important that in all of these equations, only $[\nabla f_\psi]$ and $[d^f]$ are functions of stress. Thus, the Newton–Raphson iterative process can be easily applied to solve these system equations.

9.7 System Equation Solution

In this section, we describe the Newton–Raphson iterative scheme employed to solve the system equations (Eq. 9.39). For those unfamiliar with this method, a brief description of it is given in Appendix F. During this iterative process, it is important that the true stress paths of all material points are followed accurately. In particular, the transition from elastic to nonlinear states requires careful handling. One simple strategy for accomplishing this is also described. In addition, it is convenient to provide some means of incrementing the boundary conditions automatically and, for monotonic loading, a simple scheme for this purpose is summarized here.

9.7.1 Newton–Raphson Method

The notation $\{\sigma\}_n$ and $\{x\}_n$ is used to denote the stress and boundary unknowns, respectively, at the end of the nth increment. In each increment, we use $\{\sigma\}^i$, $\{x\}^i$, and $\{\lambda\}^i$ to denote current values after the ith iteration. The residual of Eq. (9.39), following the ith iteration, can be written as

$$\{R\}^i = \{\dot{y}^f\}^i - [A^\lambda]^i \{\dot{\lambda}\}^i \tag{9.46}$$

Our objective is to reduce the residual to zero. If we force $\{R\}^{i+1}$ to be zero, we obtain

$$0 = \{R\}^{i+1} = \{R\}^i + \frac{\partial \{R\}^i}{\partial \{\lambda\}^i} \{\Delta\lambda\}$$

$$= \{R\}^i - [A^\lambda]^i \{\Delta\dot{\lambda}\} \tag{9.47}$$

where $\{\Delta\lambda\}$ are the changes in the plastic multiplier. From this equation, we obtain

$$[A^\lambda]^i \{\Delta\dot{\lambda}\} = \{R\}^i \tag{9.48}$$

Solving this equation for $\{\Delta\dot{\lambda}\}$, namely, the changes in the plastic multipliers, we obtain the corresponding changes in the boundary unknowns and stresses:

$$\{\Delta\dot{x}\} = [A^c][d^f]^i \{\Delta\dot{\lambda}\} \tag{9.49}$$

$$\{\Delta\dot{\sigma}\} = [E][d^f]^i \{\Delta\dot{\lambda}\} \tag{9.50}$$

where $[d^f]^i$ is the current value of $[d^f]$. The current values of the variables are then updated using the equations

$$\{\dot{\lambda}\}^{i+1} = \{\dot{\lambda}\}^i + \{\Delta\dot{\lambda}\} \tag{9.51}$$

$$\{x\}^{i+1} = \{x\}^i + \{\Delta\dot{x}\} \tag{9.52}$$

and

$$\{\sigma\}^{i+1} = \{\sigma\}^i + \{\Delta\dot{\sigma}\} \tag{9.53}$$

Finally, once convergence is achieved, the internal (material) variables described in Chapter Seven may be updated using the current stress state and plastic multiplier increments.

9.7.2 Transition from Elastic to Elasto–Plastic States

Numerical solution of the nonlinear system equations must perforce involve finite-sized increments. Since computational constraints demand that large increments should be employed, some care is needed to minimize the errors that arise when the stress paths followed during the iteration process violate the yield criteria. At the transition from elastic to elasto–plastic states, one method of dealing with this problem is to compute for each node, at the start of each increment, a trial stress (predictor). The trial stress $\{\sigma^t\}$ for the $(n+1)$-th increment is defined as

$$\{\sigma^t\} = \{\sigma\}_n + \{\dot{y}^e\} \tag{9.54}$$

where $\{\sigma\}_n$ is the stress at the end of the nth increment and $\{\dot{y}^e\}$ is the elastic stress increment. The yield functions $f(\sigma^t)$ and $f(\sigma_n)$ are then computed. If $f(\sigma^t) \leq 0$, the $(n+1)$-th increment is elastic and the trial stress is correct. If $f(\sigma_n) = 0$ and $f(\sigma^t) \geq 0$, the $(n+1)$-th increment is wholly elasto–plastic and the elasto–plastic algorithm is then applied. The difficulty arises when the stress state crosses from elastic to elasto–plastic, that is, when $f(\sigma_n) < 0$ and $f(\sigma^t) > 0$. This case is depicted in Fig. 9.8, in which the vector $\overrightarrow{AB}$ represents $\{\dot{y}^e\}$.

Since the elasto–plastic constitutive equations apply only when the current stress state is on the yield surface, we first scale the increment $\{\dot{y}^e\}$ to the yield surface; that is, we determine the scaling factor α, such that

$$\begin{aligned} f &= \bar{f}(\{\sigma\}_n + \alpha\{\Delta\sigma\}) - \sigma_Y \\ &= 0 \end{aligned} \tag{9.55}$$

where

$$\{\Delta\sigma\} = \{\sigma^t\} - \{\sigma\}_n \tag{9.56}$$

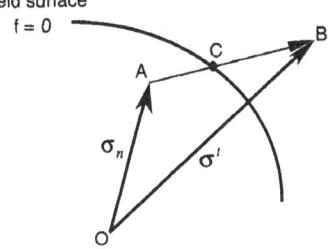

Figure 9.8: Transition from elastic to elasto–plastic states.

Determining the scaling factor α is equivalent to finding the point of intersection (denoted by the point C in Fig. 9.8) of the vector $\vec{AB}$ and the yield surface. Since the yield function is a nonlinear function of stresses, an iterative technique (e.g., Crisfield, 1997) is required. First, we assume some starting guess for α, such as

$$\alpha^0 = \frac{\sigma_Y - \bar{f}(\sigma_n)}{\bar{f}(\sigma') - \bar{f}(\sigma_n)} \tag{9.57}$$

Now, in general, after the kth iteration, Eq. (9.55) is not satisfied exactly. Forcing satisfaction of this equation for the $(k+1)$-th iteration, using the truncated Taylor series about α, produces

$$
\begin{aligned}
f^{k+1} &= f^k + \frac{\partial f}{\partial \{\sigma\}} : \frac{\partial \{\sigma\}}{\partial \alpha} \dot{\alpha} \\
&= f^k + \frac{\partial \bar{f}}{\partial \{\sigma\}} : \{\Delta\sigma\} \dot{\alpha} = 0
\end{aligned}
\tag{9.58}
$$

where $\{\sigma\} = \{\sigma\}_n + \alpha\{\Delta\sigma\}$ and $\dot{\alpha}$ is the change in α. Solving this equation, we obtain

$$\dot{\alpha} = \frac{-f^k}{\dfrac{\partial \bar{f}}{\partial \{\sigma\}} : \{\Delta\sigma\}} \tag{9.59}$$

The scaling factor can now be updated as follows:

$$\alpha^{k+1} = \alpha^k + \dot{\alpha} \tag{9.60}$$

This procedure converges very quickly; usually two or three iterations is sufficient. Indeed, for the Von Mises yield function, the starting guess α^0 is the exact solution and no iteration is required. Once the scaling factor is determined, the current stress is updated to the yield surface (at C) and the increment of stress now becomes $(1 - \alpha)\{\dot{y}^e\}$.

9.7.3 Automatic Incrementation of Boundary Loading

In many cases, it is convenient to increment the boundary loading automatically (if it increases monotonically) rather than specifying the magnitude of each increment. Provided that first yield does not occur at vanishingly small load levels, then the following procedure offers a simple and reasonably robust method for load incrementation. The first increment is taken to be that load which just produces first yield. Then, the subsequent increments are assumed to form a (diminishing) geometric series, constrained by the prescribed total number of increments and final load level. Because this simple procedure is encoded within the computer program, some of the details are given here. First, a linear solution $\{x\}$ and $\{\sigma\}$ is obtained by setting $\{\lambda\}$ to zero while prescribing the full-load level. Now, we calculate the equivalent stress for every node, and calculate the scaling factor α_f, defined by the equation

$$\alpha_f = \left(\frac{\sigma_y}{\sigma}\right)_{min} \tag{9.61}$$

where, for each node, σ_y is the initial yield stress and σ is the equivalent stress. If the initial yield stress is the same for all nodes, the search simplifies to finding the node with the highest equivalent stress. Scaling the stresses and boundary values so that this node is just at the point of yield produces

$$\begin{aligned}\{x^e\} &= \alpha_f\{x\} \\ \{\sigma^e\} &= \alpha_f\{\sigma\}\end{aligned} \tag{9.62}$$

By definition, the first increment is linear, and so these are the current boundary unknowns and stresses at the end of the first increment. Denoting, for convenience, the final boundary conditions by the vector $\{F\}$, and those imposed in the first increment by $\{F^e\}$, we see that the remaining "load" $\{F^0\}$ is

$$\{F^0\} = (1 - \alpha_f)\{F\} \tag{9.63}$$

Since the incremental loads are assumed to form a geometric series,

$$\{F^0\} = \sum_{n=1}^{N_F}\{\dot{F}\}_n \tag{9.64}$$

where $\{\dot{F}\}_n$ is the nth elasto–plastic load increment and N_F is the specified number of such increments. The ratio of successive increments is denoted by φ (where $0 < \varphi \le 1$), that is,

$$\varphi = \frac{\{\dot{F}\}_{n+1}}{\{\dot{F}\}_n} \tag{9.65}$$

For convenience, writing the first of these increments in the form

$$\{\dot{F}\}_1 = w_0\{F\} \tag{9.66}$$

we obtain

$$w_0 = 1 - \alpha_f \Big/ \sum_{n=1}^{N_F}\varphi^{n-1} \tag{9.67}$$

Because φ and N_F are input parameters, the boundary conditions for the first elasto–plastic increment can be readily obtained from Eqs. (9.66) and (9.67). The boundary conditions for the second increment and those that follow can be obtained from Eq. (9.65).

9.7.4 Summary

The complete solution process, based on the Newton–Raphson iterative algorithm described in this section, may be summarized as follows:

1. Impose load increment $\{\dot{y}^c\}$ and $\{\dot{y}^e\}$ (see Section 9.7.3).

2. Scale stresses for each node:

   $\{\sigma^t\} = \{\sigma\}_n + \{\dot{y}^e\}$

   If $f(\{\sigma^t\}) \leq 0$, then $\{\sigma\}_n = \{\sigma^t\}$

   else if $f(\sigma_n) < 0$ then:

   Determine scaling factor α (see Section 9.7.2.). Then:

   $\{\sigma\}_n = \{\sigma\}_n + \alpha\{\dot{y}^e\}$ and then set $\{\dot{y}^e\} = (1 - \alpha)\{\dot{y}^e\}$.

3. Initialize iterative variables for each node:

$$i = 0; \{\sigma\}^i = \{\sigma\}_n + \{\dot{y}^e\}; \ \{x\}^i = \{x\}_n + \{\dot{y}^c\}; \ \{\lambda\}^i = 0$$

4. Evaluate $[A^\lambda]^i$, $\{\dot{y}^f\}^i$ using Eqs. (9.40)–(9.41) and stress state $\{\sigma\}^i$.

5. Calculate residual $\{R\}^i$ using Eq. (9.46).

6. Check convergence:

   If $|\{R\}^i| = \sqrt{\{R\}^{i^T}\{R\}^i} < \varepsilon$ then go to Step 11.

7. Solve system equations for $\{\Delta\lambda\}$ using Eq. (9.48).

8. Evaluate changes in boundary unknowns and stresses using Eqs. (9.49)–(9.50).

9. Update variables for each node using Eqs. (9.51)–(9.53) and (9.43).

10. Go to Step 4 for next iteration.

11. Evaluate equivalent plastic strain increment using:

$$\{\dot{\varepsilon}^p\} = \dot{\lambda}\{\partial f/\partial\sigma\}; \ \dot{\bar{\varepsilon}}^p = c'\sqrt{\{\dot{\varepsilon}^p\}^T\{\dot{\varepsilon}^p\}}$$

and update internal variables using Eqs. (7.66) and (7.67).

12. Go to Step 1 and begin next increment.

It is crucially important to note that the matrices $[\nabla f_\psi]$ and $[d^f]$ are zero for elastic nodes. Therefore, the number of system degrees of freedom (in Eq. 9.39) is equal only to the number of yielded nodes in the current increment. This characteristic of the solution strategy saves substantial computational time over less efficient strategies. Further, the matrices $[C]$, $[A^c]$, and $[E]$ in Eqs. (9.40), (9.49), and (9.50), respectively, are constants. Once formed, they can be stored and retrieved from memory when required. Moreover, the matrices $[\nabla f_\psi]$ and $[d^f]$ need not be formed in computation, because they can be directly incorporated into $[A^\lambda]$ and $\{\dot{y}^f\}$ during the assembly process. Parenthetically, for the mixed hardening model, the stresses σ_{ij} appearing in the above formulas must be understood to be quantity $\sigma_{ij} - \rho_{ij}$.

9.8 Closure

The formulation described here is only one of several possible ways in which the nonlinear system equations may be solved. (Some others are described in Appendix F, Section F.2.) The primary motivation for developing this particular formulation was our unsatisfactory experiences with various other strategies, whose inefficiency inhibited our objective of applying the boundary element method to realistic nonlinear practical problems. Among the difficulties formerly encountered, poor convergence was a recurrent problem but excessive computational costs (per increment) and excessive demands on computer memory were also significant factors. Of course, the current formulation might also be improved in various ways (in the treatment of stress incursions beyond yield, for example) and more efficient matrix solvers might be employed, but the evidence to date suggests that our goals of computational efficiency, encompassing also algorithmic stability and accuracy, have been substantially realized. Before turning to these applications, we first describe the computer code in some detail.

CHAPTER TEN

The Elasto–Plastic Program Code

10.1 Introduction

In this chapter, we describe the extension of the Fortran computer program BEMECH to nonlinear analysis. For illustrative purposes, just four material models are encoded, but it should be a simple matter to incorporate other models if desired. All the capabilities of the linear code are preserved and thus the program can deal with plane stress, plane strain, and three-dimensional problems. The input data for the program are described in Appendix H; some typical examples are described in Chapter Eleven.

10.2 Scope of the Program

The program BEMECH is capable of analyzing linear and elasto–plastic single-region problems of arbitrary geometry in both two and three dimensions. Once the boundary unknowns are determined, interior quantities at selected points can be obtained, if desired. In both two and three dimensions, linear and quadratic isoparametric boundary elements are encoded: in the latter case, quadrilateral (rather than triangular) elements are offered. The internal cells are compatible with the boundary discretization, ranging from the 4-noded linear cell in two dimensions to the 20-noded cell in three dimensions. With regard to material models, the Tresca, Von Mises, Mohr–Coulomb, and Drucker–Prager constitutive equations are encoded. Each of these may be incorporated within isotropic, kinematic (or mixed) hardening rules, and there are no stability restrictions relating to softening. As in the linear code, any one of seven symmetry conditions may be invoked, namely, symmetry about any one, any two, or all three Cartesian planes.

10.3 Program Structure

In addition to the subroutines described in Chapter Five, a further seventeen principal subroutines are used in the nonlinear program. These are listed in Table 10.1

Table 10.1. Program units		
Name	**Section**	**Description**
CELL_BOUND	10.9	Integration over boundaries of singular cells
DF_DSIG	10.15	Calculates the derivative, $\partial f/\partial \sigma_{ij}$
DF_MATRX	10.20	Calculates matrix $\{d^f\} = [D^e]\{\partial f/\partial \sigma\}$
EVAL_KE	10.10	Evaluates values of the kernels E_{ijk} and E_{ijkl}
INPUT_NL	10.5	Reads cell nodes and plastic material properties
INT_CELL	10.7	Evaluates integrals of E_{ijk} and E_{ijkl} over cells
INTSUBC	10.11	Evaluates integrals over subcells
MATRICES	10.17	Forms matrices in terms of plastic multipliers
NL_COEFS	10.6	Evaluates coefficients of initial stresses
NL_SOLVE	10.13	Solution scheme for nonlinear system equations
PBSCOEF	10.12	Forms initial stress coefficients for boundary nodes
PL_FLOW	10.14	Evaluates equiv. stress $\bar{f}$ and vector $\partial f/\partial \sigma_{ij}$
P_M_ITER	10.16	Plastic multiplier iteration method
SIGCROSS	10.19	Finds yield point when stresses exceed yield
SIG_SCALE	10.18	Scales stresses to yield surface
SIN_CELL	10.8	Evaluates integrals over a singular cell
UPDATEV	10.21	Updates variables in iteration

and are interlinked as depicted in the flowchart in Fig. 10.1. Subsidiary flowcharts (Figs. 10.2 and 10.3) illuminate the flow paths for the branches emanating from subroutines SIN_CELL and P_M_ITER, respectively. The link for the linear analysis is quite distinct (see Fig. 5.1). The main program has been listed in Chapter Five.

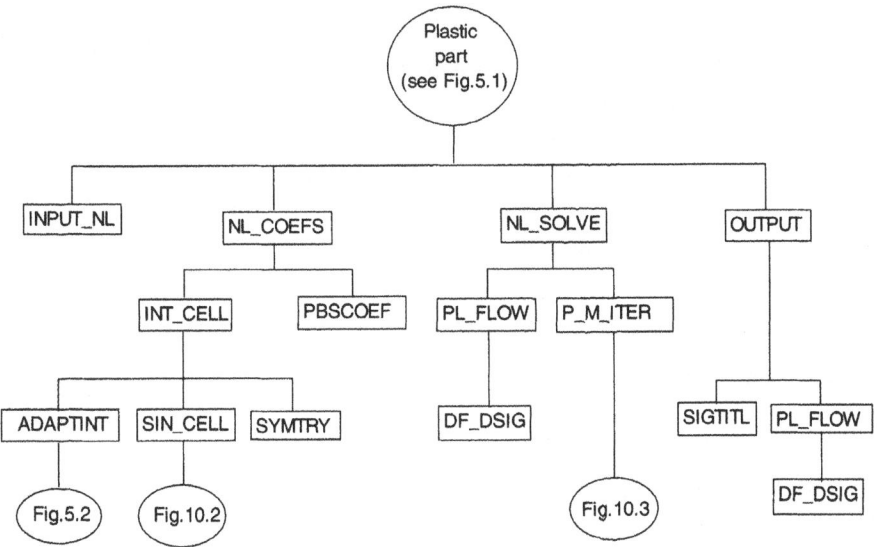

Figure 10.1: Principal flowchart for plastic analysis.

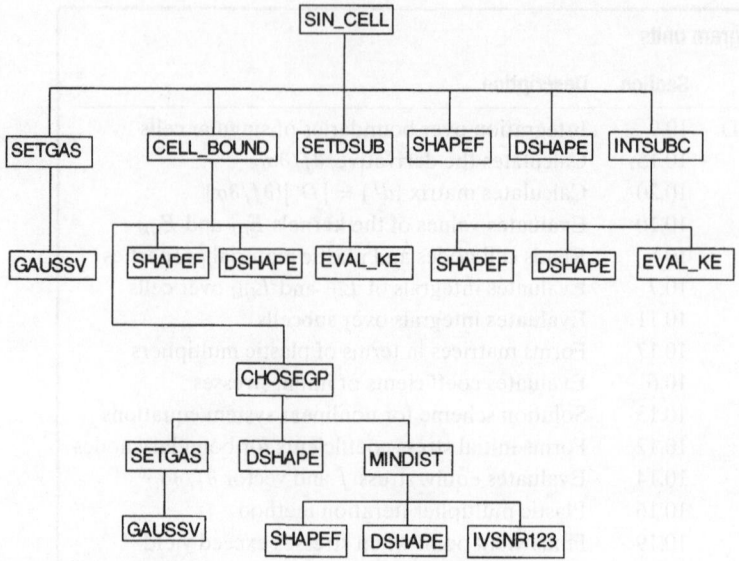

Figure 10.2: Subsidiary flowchart for subroutine SIN_CELL.

10.4 Global Variables

The following glossary includes only those global variables that were not previously defined in Chapter Five.

ALFA: Coefficient α in Drucker–Prager criterion.

BACKS: Array of back stresses.

CEQP: Factor c' in Table 7.2 for equivalent plastic strain.

CEQS: Factor γ and χ in Table 7.1 for equivalent stress.

CJUP: Coefficient for stress jump terms (free terms).

CLAME: Coefficient, $\nu/(1 - 2\nu)$.

DFDS: Array containing $\partial f/\partial\sigma$.

EM: Hardening scale factor between 0 and 1 (0 = kinematic to 1 = isotropic).

EPSTN: Vector of equivalent plastic strain.

FINCR: Ratio of two adjacent load increments.

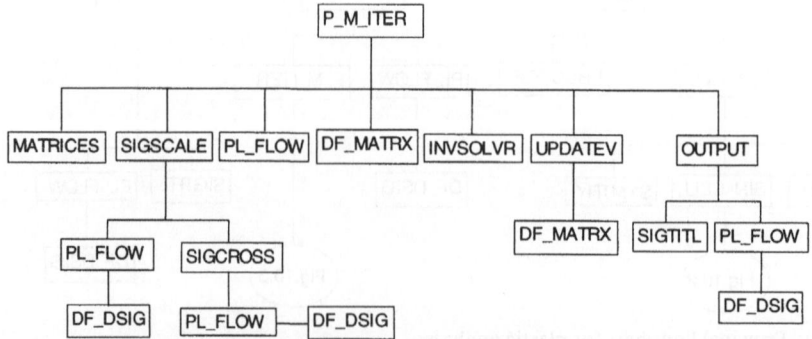

Figure 10.3: Subsidiary flowchart for subroutine P_M_ITER.

FRICT: Friction angle ϕ (degrees) for Mohr–Coulomb and Drucker–Prager criteria.

HARDI: Isotropic hardening modulus H'.

HARDK: Kinematic hardening modulus $\overline{H}$.

HARDS: Equivalent mixed hardening modulus H^*.

INY: Vector that restores the stress vector STRES to tensor form. In three dimensions:

$$\begin{bmatrix} \text{STRES}(1) & \text{STRES}(4) & \text{STRES}(6) \\ \text{STRES}(4) & \text{STRES}(2) & \text{STRES}(5) \\ \text{STRES}(6) & \text{STRES}(5) & \text{STRES}(3) \end{bmatrix}$$

LNDC: Array of cell nodal connectivity; numbering as in Appendix B.

MITER: Maximum permitted number of iterations for each increment.

MULTP: Number of traction-discontinuous nodes.

NBTP: Number of basic boundary nodes and interior nodes. Basic boundary nodes: $1 - \text{NBP}$; interior nodes: $\text{NBP} + 1$ to NBTP; additional nodes (traction discontinuous): $\text{NBTP} + 1$ to NTP.

NCRIT: Flag: 1 = Tresca; 2 = Von Mises; 3 = Mohr–Coulomb; 4 = Drucker–Prager.

NINCS: Number of load increments.

NNOD: Vector of yielded node numbers.

NODC: Number of cell nodes.

NOUT1:   Output flag for each increment.

NOUT2:   Output flag for final results.

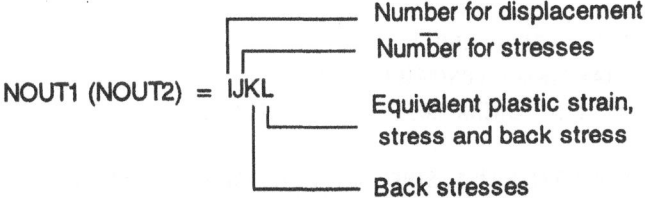

The values of I, J, K, and L can vary from 0 to 9. Nonzero values in NOUT2 switch on printing. So do those for NOUT1, but only if the increment number is also an integer multiple of the non zero value.

PKERN: Array of kernels E_{ijk} and E_{ijkl}.

RE: Vector used to compute stress–strain products.

ROOT3: $\sqrt{3}$.

TOLER: Iteration tolerance.

UNIAX: Uniaxial initial yield stress σ_y for Tresca and Von Mises criteria and cohesion c for Mohr–Coulomb and Drucker–Prager criteria

10.5 Subroutine INPUT_NL

This routine reads cell nodes, elasto–plastic material properties, and parameters controlling the solution process. The description of input variables is listed in

Appendix H. Since stresses are normalized by the factor $2G$, the material parameters (including the hardening parameters) are normalized too.

Local variables

M: Cell number.
NODC: Number of cell nodes (4, 8, or 20).

```
      SUBROUTINE INPUT_NL(NODC)
      USE FIXED_VALUES; USE VARY_ARRAYS
      WRITE(7,'(///25X,"PLASTIC PART :"//4X,"CELL",              &
     & 19X,"...NODES...")')
      DO 10 ICELL=1,NCELL    ! Read nodes of interior cells
      READ(5'*)M,(LNDC(INODE,M),INODE=1,NODC)
10    WRITE(7,'(1X,11I6/7X,10I6)')M,(LNDC(INODE,M),INODE=1,NODC)
      READ(5'*)NCRIT,UNIAX,FRICT,HARDI,HARDK,EM  ! Material properties
      WRITE(7,20)NCRIT,UNIAX,FRICT,HARDI,HARDK,EM
20    FORMAT(/1X,'NCRIT',6X,'UNIAX',8X,'FRICT',8X,'HARDI',8X,'HARDK',  &
     & 9X,'EM'/1X,I3,3X,1P5E13.5)
!                  Input incremental information
      READ(5,*)NINCS,FINCR,MITER,TOLER,NOUT1,NOUT2
      WRITE(7,30)NINCS,FINCR,MITER,TOLER,NOUT1,NOUT2
!        Set up nonlinear parameters  (some normalized by 2G)
      ROOT3=SQRT(3.); CLAME=PR/PR2
      CJUP=NDIM*NDIM-2.-(NDIM*NDIM-4.)*PR
      FRICT=FRICT*0.017453292; SINF=SIN(FRICT)
      IF(NCRIT.EQ.4) ALFA=2.*SINF/(ROOT3*(3.-SINF))
      IF(NCRIT.EQ.3) CEQS=0.5*(1.-SINF)
      IF(NCRIT.EQ.4) CEQS=-ALFA+1./ROOT3
      IF(NCRIT.EQ.3) UNIAX=UNIAX*COS(FRICT)/CEQS
      IF(NCRIT.EQ.4) UNIAX=(6.*UNIAX*COS(FRICT)/(ROOT3*(3.-SINF)))/CEQS
      CEQP=SQRT(2./3.)
      IF(NCRIT.EQ.3) CEQP=SQRT(2.)*(1.-SINF)/SQRT(3.-2.*SINF+3.*SINF**2)
      IF(NCRIT.EQ.4) CEQP=(1./ROOT3-ALFA)/SQRT(3.*ALFA*ALFA+0.5)
      UNIAX=UNIAX/2./G; HARDI=EM*HARDI/2./G; HARDK=(1.-EM)*HARDK/2./G
      HARDS=HARDI+HARDK
30    FORMAT(/1X,'NINCS   FINCR    MITER    TOLER     NOUT1',         &
     &          '   NOUT2',/1X,I4,4X,F8.6,I7,1PE14.4,I7,I10)
      END SUBROUTINE INPUT_NL
```

10.6 Subroutine NL_COEFS

This routine calls routines that evaluate initial stress coefficients for both boundary and interior nodes (refer to Section 9.6.1).

Local variables

COEF: Array containing matrices $[E^b]$ and $[E^\sigma]$ (Eqs. 9.22–9.24).
NNOD: Number of elements meeting at a node.

```
      SUBROUTINE NL_COEFS(NODE,NODC)
      USE VARY_ARRAYS; USE FIXED_VALUES
      WRITE(*,'(//10x," Forming PLASTIC Coefficient Matrices")')
      IF(NDIM.EQ.2) NGSIN(3)=1  ! Only two components for 2D problems
!         Compute coefficients of initial stress for displacements
      CALL INT_CELL(NODE,NODC,4,NDIM,1,NBTP,COEF,51)
!     Initial stress coefficients for stresses (boundary nodes)
      CALL PBSCOEF(1,NBP,NNOD)
!     Initial stress coefficients for stresses (interior nodes)
      CALL INT_CELL(NODE,NODC,5,NSIG,NBP+1,NBTP,COEF,53)
!     Initial stress coefficients for stresses (additional nodes)
      CALL PBSCOEF(NBTP+1,NTP,NNOD)
      END SUBROUTINE NL_COEFS
```

10.7 Subroutine INT_CELL

This routine evaluates domain integrals (refer to Sections 9.3 and 9.4) for displacements at boundary and interior nodes and for stresses at interior nodes.

Local variables

COEF: Array containing matrices $[E^b]$ and $[E^\sigma]$.
MSUB: Number of cell sides (4 or 6).
NBEG: First node number.
NCS: Node number of singular cell.
NEND: Last node number.
NROW: $= 3, 4$, or $6(E_{ijkl})$, for plane stress, plane strain, or three dimensions.
NTU: Flag for integrals: $4 =$ kernel E_{ijk}, $5 =$ kernel E_{ijkl}.
PKERN: Array containing integrals of E_{ijk} or E_{ijkl}.
PSYM: Vector of symmetry multipliers (± 1) for initial stresses.
RLEVL: Singularity order for adaptive Gauss integration.

```
      SUBROUTINE INT_CELL(NODE,NODC,NTU,NROW,NBEG,NEND,COEF,NFL)
      USE VARY_ARRAYS,ONLY:CD,KBU,LNDC; USE FIXED_VALUES
      REAL COEF(NROW,NTSF),CP(3)
      MSUB=NDIM/3*2+4; K0=NDIM/3*12
      IF(NAUTO.LE.0) RLEVL=NDIM-5+NTU  ! =1(2)Eijk & 2(3)Eijkl in 2(3)D
      DO 90 IPOIN=NBEG,NEND            ! Loop over nodes
      WRITE(*,10)IPOIN,NEND,NTU
10    FORMAT(/15X,'NODE == ',I4,' /',I4,5X,'FOR NTU ==',I3)
      CP(1:NDIM)=CD(1:NDIM,IPOIN); COEF(1:NROW,1:NTSF)=0.
      DO 50 ICELL=1,NCELL             ! Loop over interior cells
      PKERN(1:NROW,1:NSIG*NODC)=0.; NCS=0
      DO 20 ID=1,NODC; ND(ID)=ID
      NW=LNDC(ID,ICELL); IF(NW.GT.NBTP) NW=KBU(NW)
```

```
        IF(NW.EQ.IPOIN) NCS=ID          ! Choose singular node
        LND=LNDC(ID,ICELL)
20      CK(1:NDIM,ID)=CD(1:NDIM,LND)  ! cell nodal coordinates
        PSYM(1:NSIG)=1.                 ! Assign values for original cell
        DO 30 ISY=1,NCSYM               ! Consider symmetry condition
        IF(NCS.EQ.0) CALL ADAPTINT(NDIM,NODC,0,CP,NTU)  ! Nonsingular cell
        IF(NCS.NE.0) CALL SIN_CELL(NODC,NODE,NCS,NTU,CP,MSUB,K0)
30      IF(ISY.LT.NCSYM) CALL SYMTRY(NDIM,NODC,NSYM(ISY+NSYM0),NCS,0,ND,  &
        & CSYM,CP,CK,PSYM,NSWP,2)
!                       Assembling initial stress coefficients
        DO 40 ID=1,NODC
        NP=NSIG*(LNDC(ID,ICELL)-1); JP=NSIG*(ID-1)
        DO 40 I=1,NROW  !NROW (=3 if NTU=4=Eijk)or(= NSIG if NTU=5=Eijkl)
        MP=I; IF(NTU.EQ.5) MP=INX(I)
        DO 40 J=1,NSIG
40      COEF(I,NP+J)=COEF(I,NP+J)+RE(J)*PKERN(MP,JP+INX(J))
50      CONTINUE
        IF(NTU.EQ.4) GOTO 80
!                       Deal with free terms for interior stresses
        IP=NSIG*(IPOIN-1); NP=0
        DO 70 K=1,NDIM; DO 70 L=K,NDIM; NP=NP+1; MP=0
        DO 60 I=1,NDIM; DO 60 J=I,NDIM; MP=MP+1
        TERM=-1./(NDIM*(NDIM+2.)*PR1)*(2.-DLT(K,L))*(0.5*CJUP*(DLT(I,K)*  &
        &  DLT(J,L)+DLT(I,L)*DLT(J,K))+(1.-GAM*PR)*DLT(I,J)*DLT(K,L))
60      COEF(INX(MP),IP+INX(NP))=COEF(INX(MP),IP+INX(NP))+TERM
        IF(NSIG.EQ.4) COEF(4,IP+INX(NP))=COEF(4,IP+INX(NP))               &
        &            +PR/(2.*PR1)*DLT(K,L)              !
70      CONTINUE
        IF(NSIG.EQ.4) COEF(4,IP+4)=COEF(4,IP+4)-1.
!            Store initial stress coefficients on disk
80      DO I=1,NROW; WRITE(NFL)(COEF(I,J),J=1,NTSF); ENDDO
90      CONTINUE
        END SUBROUTINE INT_CELL
```

10.8 Subroutine SIN_CELL

This routine evaluates the integral (Eq. 9.10) of the initial strain kernel over a singular cell (the source point is coincident with one of the nodes of the cell). The cell subdivision technique, described in Section 9.3, is employed to evaluate the domain integral.

Local variables

CP: Vector of coordinates of the source point.

CSID: Array of nodal coordinates of cell side.

CSUB: Array of intrinsic coordinates of a degenerate subcell.

FJCBL: Local Jacobian of a subcell.

IPN: Position of a node in the array NODEF.

ISO: Location of singular node in array CORDL.

K0:   Used to determine the position of a node in the array NODEF.

NCOR: Number of corner nodes in the cell.

NSB: Vector (flag): 0 = degenerate side, 1 = normal side.

NSC: Number of nodes on cell side (2, 3, 4, or 8).

XIS: Vector of intrinsic coordinates of the singular node.

```
      SUBROUTINE SIN_CELL(NODC,NODE,NCS,NTU,CP,MSUB,K0)
      USE FIXED_VALUES
      DIMENSION CP(3),XI(3),NSB(MSUB),XIS(3),CSID(3,8)
      ISN=NCS; NSC=3+NDIM/3*5
!                      Set up degenerate cell flag
      DO 50 ISUB=1,MSUB
      NSB(ISUB)=1        ! =1 side can form a degenerate subcell
      DO 40 ID=1,NODE
      IPN=NSC*(ISUB-1)+ID+K0; IF(NODEF(IPN).NE.ISN) GO TO 40
      NSB(ISUB)=0        ! =0 side can not form a subcell
      GOTO 50
40    CONTINUE
50    CONTINUE
      NCOR=2**NDIM; IF(NTU.EQ.4) ISN=0
      ISO=3*(NCS+(3-NDIM)*NCS/5*4-1); XIS(1:3)=CORDL(ISO+1:ISO+3)
      DO 90 ISUB=1,MSUB; IF(NSB(ISUB).EQ.0) GO TO 90
!              Evaluate the second integral
      IF(NTU.EQ.5) THEN
!           Set up global coordinates of a side (surface)
      IPN=NSC*(ISUB-1)+K0
      DO ID=1,NODE; CSID(1:NDIM,ID)=CK(1:NDIM,NODEF(IPN+ID)); ENDDO
!             Perform integration over a side (surface)
      CALL CELL_BOUND(NTU,ISN,CP,NODE,CSID)
      ENDIF
!             Evaluate the first integral
!                  Carry out subcell division
      CALL SETDSUB(NDIM,NCOR,ISUB,NSC,CORDL,48,NODEF(1+K0),CSUB,XIS)
!        Set up Gauss points for isolated integrals
      CALL SETGAS(NGSIN(1),NGSIN(2),NGSIN(3))
!                Integrating the isolated cell
      DO IG=1,NGSS; IP=NDIM*(IG-1)+1
!            Calculate intrinsic coordinate XI in original cell
       CALL SHAPEF(NDIM,NDIM,NCOR,SHAP,GP(IP),CSUB,CPO,XI,R2,CORDL)
!             Calculate local Jacobian of a subcell
       CALL DSHAPE(NDIM,NDIM,NCOR,GP(IP),CSUB,COSB,FJCBL,DN,GD,CORDL)
!            Evaluate integrals over the subcell
      CALL INTSUBC(NODC,ISN,NTU,CP,XI,IG,FJCBL)
      ENDDO
90    CONTINUE
      END SUBROUTINE SIN_CELL
```

10.9 Subroutine CELL_BOUND

This routine evaluates the second integral in Eq. (9.10) over the boundaries of a singular cell. This is nonsingular, but if p is very close to the boundary, then an element subdivision technique similar to that used in subroutine ADAPTINT (and described in Section 4.5) is employed.

Local variables

AL:	Vector of global lengths of subelements along intrinsic coordinate directions.
CSID:	Array of global coordinates of nodes over a cell boundary.
DGS:	Vector of intrinsic coordinate lengths of subelements.
DISL:	Distance from XIC to subelement.
FJACBL:	Local Jacobian of a subelement.
MGAUS:	Maximum Gauss order.
NGASW:	Second Gauss integration order for surface integrals. In two dimensions, no second integration is needed.
NCS:	Node number of the singular cell.
R2:	r^2.
RI:	Vector r_i.
RN:	$r_m n_m$.
WX/Y:	Gauss weights for first/second integration.
XI:	Vector of coordinates of Gauss points.
XIC:	Vector of intrinsic coordinates of the proximate point.

```
       SUBROUTINE CELL_BOUND(NTU,NCS,CP,NODE,CSID)
       USE FIXED_VALUES
       REAL CP(3),CSID(3,NODE),RI(3),XI(3),XIC(3),DGS(3),AL(3)
!      Determine XIC, DGS and necessary number of subelements
       MGAUS=10    ! If MGAUS increased, then change subroutine GAUSSV
       CALL CHOSEGP(NBDM,NODE,CP,CSID,DGS,XIC,AL,DISL,1,MGAUS)
!
       IF(MITER.NE.1) GOTO 20 !If subdivision used, delete this
!                    No subdivision
       IF(NSUB(1)+NSUB(2).GT.2) GOTO 30
20     DO IG=1,NGSS; IP=NBDM*(IG-1)+1
         CALL SHAPEF(NDIM,NBDM,NODE,SHAP,GP(IP),CSID,CP,RI,R2,CORDL)
         CALL DSHAPE(NDIM,NBDM,NODE,GP(IP),CSID,COSB,FJCB,DN,GD,CORDL)
         RN=DOT_PRODUCT(RI(1:NDIM),COSB(1:NDIM))
         WJCB=FJCB*GW(IG)*RN
         IF(NTU.EQ.5) WJCB=WJCB*ALOG(SQRT(R2))
         CALL EVAL_KE(RI,R2,WJCB,1,NTU,NCS)
       ENDDO
       RETURN
!                    Subdivision cases
30     NCOR=2**NBDM             ! Number of element corner nodes
```

```
!           Detm. intrinsic coords. in 2nd direction for 1st subelement
      DO ID=1,NCOR; CSUB(2,ID)=-1.+(CORDL(3*ID-1)+1.)/NSUB(2); ENDDO
!           Loop of subelement along the 2nd intrinsic direction
      DO 60 ISUB2=1,NSUB(2)  ! No. of subcells in 2nd directn(=1 in 2D)
!           Determine coordinates in the 1st direction for subelements
      DO ID=1,NCOR; CSUB(1,ID)=-1.+(CORDL(3*ID-2)+1.)/NSUB(1); ENDDO
!           Loop of subelement along the first intrinsic direction
      DO 50 ISUB1=1,NSUB(1)    ! No. of subcells in first direction
!           Find local proximate point XI within the subelement
      CALL MINDIST(NDIM,NBDM,NCOR,XIC,CSUB,1.,XI,DISL)
!         Find intrinsic coordinates RI for  XI in the original element
      CALL SHAPEF(NDIM,NBDM,NCOR,SHAP,XI,CSUB,CP0,RI,R2,CORDL)
!         Find distance SQRT(R2)from source point to subelement
      CALL SHAPEF(NDIM,NBDM,NODE,SHAP,RI,CSID,CP,XI,R2,CORDL)
!              Calculate Gauss orders for the subelement
      CALL CHOSEGP(NBDM,NODE,CP,CSID,DGS,XI,AL,SQRT(R2),0,MGAUS)
!                  Integrating over the subelement
      DO 40 IG=1,NGSS; IP=NBDM*(IG-1)+1
!           Find global intrinsic coordinates XI for the Gauss point IG
      CALL SHAPEF(NDIM,NBDM,NCOR,SHAP,GP(IP),CSUB,CP0,XI,R2,CORDL)
!           Find local Jacobian FJCBL for the subelement (cell)
      CALL DSHAPE(NDIM,NBDM,NCOR,GP(IP),CSUB,COSB,FJCBL,DN,GD,CORDL)
!                      Evaluate boundary integrals
       CALL SHAPEF(NDIM,NBDM,NODE,SHAP,XI,CSID,CP,RI,R2,CORDL)
       CALL DSHAPE(NDIM,NBDM,NODE,XI,CSID,COSB,FJCB,DN,GD,CORDL)
       RN=DOT_PRODUCT(RI(1:NDIM),COSB(1:NDIM))
       WJCB=FJCBL*FJCB*GW(IG)*RN
       IF(NTU.EQ.5) WJCB=WJCB*ALOG(SQRT(R2))
       CALL EVAL_KE(RI,R2,WJCB,1,NTU,NCS)
40    CONTINUE
!           Compute intrinsic coordinates for next subelement
50    CSUB(1,1:NCOR)=CSUB(1,1:NCOR)+DGS(1)  ! First intrinsic direction
60    CSUB(2,1:NCOR)=CSUB(2,1:NCOR)+DGS(2)  ! Second intrinsic direction
      END SUBROUTINE CELL_BOUND
```

10.10 Subroutine EVAL_KE

This routine evaluates the kernels E_{ijk} and E_{ijkl} (from Eqs. 8.13 and 8.24, respectively).

Local variables

GAM:	Parameter γ.
NBDM:	Parameter α.
NCS:	Node number of the singular cell.
NDIM:	Parameter β.

NODD: Flag: 1 = dealing with second integral in Eq. (9.10); NODC = dealing with domain integrals.

NTU: Flag : $4 = E_{ijk}, 5 = E_{ijkl}$.

RI: Vector r_i.

SHP: 1 = boundary integrals; N_α = nonsingular; $N_\alpha - 1$ = singular.

VKE: Array containing E_{ijk} and E_{ijkl}.

Z: Vector $r_{,i} = r_{i/r}$.

```
      SUBROUTINE EVAL_KE(RI,R2,WJCB,NODD,NTU,NCS)
      USE FIXED_VALUES; REAL VKE(3,3,3,3),RI(3),Z(3)
      CONP=CON*WJCB/(SQRT(R2))**NBDM; Z(1:NDIM)=RI(1:NDIM)/SQRT(R2)
      DO 40 K=1,NDIM; DO 40 L=K,NDIM; DO 40 I=1,NDIM
      DO 30 J=I,NDIM; GOTO (10,20),NTU-3
!         Form initial stress coefficient Eikl for displacement
10    TM1=PR2*(DLT(I,K)*Z(L)+DLT(I,L)*Z(K))-DLT(K,L)*Z(I))
      VKE(I,I,K,L)=CONP*(TM1+NDIM*Z(I)*Z(K)*Z(L))
      GOTO 40
!         Form initial stress coefficient Eijkl for interior stress
20    TM1=PR2*(DLT(I,K)*DLT(J,L)+DLT(I,L)*DLT(J,K)-DLT(I,J)*DLT(K,L)    &
   &     +NDIM*DLT(I,J)*Z(K)*Z(L))+NDIM*DLT(K,L)*Z(I)*Z(J)
      TM2=NDIM*PR*(DLT(I,K)*Z(J)*Z(L)+DLT(J,K)*Z(I)*Z(L)+DLT(I,L)*    &
   &     Z(J)*Z(K)+DLT(J,L)*Z(I)*Z(K))-NDIM*GAM*Z(I)*Z(J)*Z(K)*Z(L)
30    VKE(I,J,K,L)=-CONP/SQRT(R2)*(TM1+TM2)
40    CONTINUE
!                     Allocate Eikl or Eijkl to cell matrix
      DO 70 ID=1,NODD
      IF(NODD.EQ.1) THEN
       IP=NSIG*(ND(NCS)-1); SHP=1.        ! For second integral
      ELSE
       IP=NSIG*(ND(ID)-1); SHP=SHAP(ID)  ! For nonsingular node
       IF(ID.EQ.NCS) SHP=SHAP(ID)-1.     ! For singular node
      ENDIF
      NP=0       ! Index for 1D storage of cell nodes
      DO 70 K=1,NDIM; DO 70 L=K,NDIM; NP=NP+1
      MP=0       ! Index for 1D storage of nodal components
      DO 60 I=1,NDIM; DO 50 J=I,NDIM
      MP=MP+1            ! Form cell matrix
      PKERN(MP,IP+NP)=PKERN(MP,IP+NP)+VKE(I,J,K,L)*SHP*PSYM(NP)
50    IF(NTU.EQ.4) GOTO 60       ! Skip index j for Eikl
60    CONTINUE
!                     Eijkl for plane-strain case
      IF(NSIG.EQ.4.AND.NTU.EQ.5) PKERN(4,IP+NP)=PKERN(4,IP+NP)        &
   &              +PR*(VKE(1,1,K,L)+VKE(2,2,K,L))*SHP*PSYM(NP)
70    CONTINUE
      END SUBROUTINE EVAL_KE
```

10.11 Subroutine INTSUBC

This routine evaluates domain integrals over a subcell.

Local variables

CP: Coordinates of source point.
FJCB: Global Jacobian J_c.
FJCBL: Local Jacobian J_c^s.
RI: Vector r_i.
R2: r^2.
XI: Vector of global intrinsic coordinates $\xi_1, \xi_2(\xi_3)$.

```
      SUBROUTINE INTSUBC(NODC,NCS,NTU,CP,XI,IG,FJCBL)
      USE FIXED_VALUES; REAL CP(3),XI(3),RI(3)
!                          Calculate Ri and R**2
      CALL SHAPEF(NDIM,NDIM,NODC,SHAP,XI,CK,CP,RI,R2,CORDL)
!                          Calculate global Jacobian FJCB
      CALL DSHAPE(NDIM,NDIM,NODC,XI,CK,COSB,FJCB,DN,GD,CORDL)
      WJCB=FJCB*FJCBL*GW(IG)
!                          Evaluate kernels Eijk and Eijkl
      CALL EVAL_KE(RI,R2,WJCB,NODC,NTU,NCS)
      END SUBROUTINE INTSUBC
```

10.12 Subroutine PBSCOEF

This routine evaluates the initial stress coefficients C_{mnkl} (refer to Section 9.5) that are used to determine boundary stresses using the traction-recovery method. The results are stored on Channel 53.

Local variables

COSL: Array of direction cosines (L_{ij}) of local coordinate axes (calculated and stored on Channel 55 in subroutine BDSTRS).
IP1:   First node number.
IP2:   Last node number.
NNOD: Vector of numbers of elements meeting at nodes.
PCOEF: Array containing C_{mnkl}.

```
      SUBROUTINE PBSCOEF(IP1,IP2,NNOD)
      USE FIXED_VALUES
      DIMENSION NNOD(NTP),PCOEF(NSIG,NSIG),COSL(NDIM,NDIM)
      IF(IP1.EQ.1) REWIND(55)
      DO 60 IP=IP1,IP2; IF(NNOD(IP).EQ.0) GOTO 60
      PCOEF=0.
      DO 50 IE=1,NNOD(IP); READ(55)COSL ! Retrieve Lij
      MP=0
      DO 30 M=1,NDIM; DO 30 N=M,NDIM; MP=MP+1
```

```
         NTM3=0
         DO 30 K=1,NDIM; DO 30 L=K,NDIM; NTM3=NTM3+1
         IF(NDIM.EQ.3) GOTO 10
!
         TRM=COSL(1,M)*COSL(1,N)*(DLT(K,L)-1./PR1*COSL(2,K)*COSL(2,L))
         GOTO 20
10       TEM=0.          ! TEM contains the first term
         DO I=1,NDIM
          TEM=TEM+COSL(I,M)*COSL(I,N)*COSL(I,K)*COSL(I,L) !Lim Lin Lik  Lil
         ENDDO
         TRM=TEM-PR/PR1*DLT(M,N)*COSL(3,K)*COSL(3,L)-                    &
     &       PR2/PR1*COSL(3,M)*COSL(3,N)*COSL(3,K)*COSL(3,L)+           &
     &       (COSL(1,M)*COSL(2,N)+COSL(2,M)*COSL(1,N))*                &
     &       0.5*(COSL(1,K)*COSL(2,L)+COSL(2,K)*COSL(1,L))          !
!        The following multiplier deals with shear stress symmetry
20       PCOEF(MP,NTM3)=PCOEF(MP,NTM3)-TRM*(2.-DLT(K,L))
30       CONTINUE
         IF(NSIG.NE.4) GOTO 50; N=0
         DO 40 K=1,NDIM; DO 40 L=K,NDIM; N=N+1
!
40       PCOEF(4,N)=PCOEF(4,N)+PR/PR1*COSL(2,K)*COSL(2,L)*(2.-DLT(K,L))
         PCOEF(4,4)=-1.       ! For z-direction stress in plane strain
50       CONTINUE
!          Set off-diagonal blocks to zero
         DO L=1,NSIG; WRITE(53)(0.,M=1,NSIG*(IP-1)),(PCOEF(INX(L),      &
     &      INX(M))/NNOD(IP),M=1,NSIG),(0.,M=NSIG*IP+1,NSIG*NTP)
         ENDDO
60       CONTINUE
         END SUBROUTINE PBSCOEF
```

10.13 Subroutine NL_SOLVE

This routine automatically determines the incremental loads using the algorithm described in Section 9.7.3. For the first increment, the data are stored on Channel 55. Finally, the nonlinear system equations are solved employing the Newton–Raphson iterative scheme.

Local variables

EFSIG: Maximum equivalent elastic stress.

FINCR: Loading factor φ.

FMULT: φ^n

FOVER: Load to be divided into increments $(F - F^e)$.

FSCAL: Scaling factor α_f.

FSUM: $\sum_{n=1}^{N_F} \varphi^{n-1}$.

INCRD: Default number of increments.

TOL: Tolerance for scaling stresses to yield surface.

YIELD: Equivalent uniaxial stress (Table 7.1).

```
     SUBROUTINE NL_SOLVE(NODE,EFSIG)
     USE VARY_ARRAYS; USE FIXED_VALUES
     WRITE(*,'(//11x,"Solving  Nonlinear System Equations")')
!                      Find maximum equivalent stress
     EFSIG=0.    ! Initial value of maximum equivalent stress
     DO IP=1,NTP
      CALL PL_FLOW(ESTRS(1,IP),YIELD,STRES,0)
      IF(EFSIG.LT.YIELD) EFSIG=YIELD
     ENDDO
     WRITE(7,10)2.*G*UNIAX,2.*G*EFSIG
     IF(EFSIG.LE.UNIAX) STOP   ! All nodes are in elastic state
10   FORMAT(/1X,'STRESS LIMIT ==',1PE13.6,5X,                        &
     & 'MAXIMUM EQU. STRESS  ==',1PE13.6)
!         Determine default number of load increments
     INCRD=6*INT(SQRT(40.*(EFSIG/UNIAX-1.)+1.))
     WRITE(7,'(/7X,"DEFAULT INCREMENT ==",I5)')INCRD
!       Only when input value of NINCS=0 does default take effect
     IF(NINCS.EQ.0) NINCS=INCRD
     WRITE(7,'(7X,"ACTUAL INCREMENTS ==",I5)') NINCS
     TOL=0.001*UNIAX; FMULT=1.; FSUM=1.
     FSCAL=UNIAX/EFSIG              !
!                 Determine loading factor
     DO I=1,IABS(NINCS)-1
       FMULT=FMULT*FINCR         ! Forming phi**(n-1)
       FSUM=FSUM+FMULT           ! Sum of phi**(n-1)  (n=1-Nf)
     ENDDO
     WOVER=(1.-FSCAL)/FSUM       !
     REWIND(55)                  ! contains the first increment data
!                        store boundary unknowns
     WRITE(55)(X(I)*WOVER,I=1,NTF)
!                        store stresses
     WRITE(55)((ESTRS(I,IP)*WOVER,I=1,NSIG),IP=1,NTP)
!                        store initial boundary unknowns
     X(1:NTF)=X(1:NTF)*FSCAL       !
!                        store initial values of stresses
     ESTRS(1:NSIG,1:NTP)=ESTRS(1:NSIG,1:NTP)*FSCAL   !
!   Solve nonlinear system equation
     CALL P_M_ITER(NODE,TOL,FSCAL,WOVER)
     END SUBROUTINE NL_SOLVE
```

10.14 Subroutine PL_FLOW

This routine evaluates the equivalent stress $\bar{f}$ and calls DF_DSIG, which calculates $\partial f / \partial \sigma_{ij}$. For three dimensions, the relevant equations may be found in Section 7.2.1. However, in plane stress, some care is needed to accommodate the fact that although the stress component σ_{33} is zero, the deviatoric stress σ'_{33} is not zero. It can be determined from the equation

$$\sigma'_{33} = -\frac{1}{3}\sigma_{kk} \tag{10.1}$$

For both plane stress and plane strain problems, the second and third deviatoric stress invariants J_2 and J_3 can be obtained from the equations

$$J_2 = \frac{1}{2}\sigma'_{ij}\sigma'_{ij} + \frac{1}{2}\sigma'_{33}\sigma'_{33} \tag{10.2}$$

and

$$J_3 = \frac{1}{3}\sigma'_{ij}\sigma'_{jk}\sigma'_{ki} + \frac{1}{3}\sigma'_{33}\sigma'_{33}\sigma'_{33} \tag{10.3}$$

where the subscripts i, j, and k range from 1 to 2 only.

Local variables

DEVIA:　　　Array of deviatoric stresses σ'_{ij}.
NFL:　　　　Flag: $0 = $ calculate $\bar{f}$; $\neq 0$ calculate $\bar{f}$ and $\partial f / \partial \sigma_{ij}$.
ROOT3:　　　$\sqrt{3}$.
SIGMA:　　　Vector of stresses σ_{ij}.
SINT3:　　　$\mathrm{Sin}3\theta$.
SMEAN:　　　Mean stress, $\sigma_{kk}/3$.
SQUJ2:　　　$\sqrt{J_2}$.
THETA:　　　Angle θ.
VARJ2:　　　Second deviatoric stress invariant J_2.
VARJ3:　　　Third deviatoric stress invariant J_3.
YIELD:　　　$\bar{f}$ (from Table 7.1).

```
      SUBROUTINE PL_FLOW(SIGMA,YIELD,DFDS,NFL)
      USE FIXED_VALUES; REAL SIGMA(NSIG),DFDS(NSIG),DEVIA(3,3)
!                   Evaluate mean stress
      SMEAN=0.; DO I=1,NDIM; SMEAN=SMEAN+SIGMA(I)/3.; ENDDO
      IF(NSIG.EQ.4) SMEAN=SMEAN+SIGMA(4)/3.     ! Plane-strain case
!                   Evaluate deviatoric stresses
      K=0; DO 20 I=1,NDIM; DO 20 J=1,NDIM; K=K+1
20    DEVIA(I,J)=SIGMA(INY(K))-SMEAN*DLT(I,J)          !
!        Find second and third invariants of deviatoric stresses
      VARJ2=0.; VARJ3=0.
      DO 30 I=1,NDIM; DO 30 J=1,NDIM
      VARJ2=VARJ2+0.5*DEVIA(I,J)*DEVIA(I,J)            !
      DO 30 K=1,NDIM
```

```
30      VARJ3=VARJ3+DEVIA(I,J)*DEVIA(J,K)*DEVIA(K, I)/3. !
        IF(NSIG.EQ.6) GOTO 35
!               For plane stress/strain:
        IF(NSIG.EQ.3) DEVIA(3,3)=-SMEAN               !
        IF(NSIG.EQ.4) DEVIA(3,3)=SIGMA(4)-SMEAN       !
        VARJ2=VARJ2+0.5*DEVIA(3,3)*DEVIA(3,3)         !
        VARJ3=VARJ3+DEVIA(3,3)**3/3.                  !
35      SQUJ2=SQRT(VARJ2); IF(SQUJ2.LT.1.E-6) GOTO    40
        SINT3=-3.*ROOT3*VARJ3/(2.*VARJ2*SQUJ2)        !
        IF(ABS(SINT3).GT.1.) SINT3=SIGN(1.,SINT3)
        GO TO 50
40      SINT3=0.0
50      THETA=ASIN(SINT3)/3.      !
!                       Find equivalent stress for four criteria
        SELECT CASE(NCRIT)
          CASE(1); YIELD=2.*COS(THETA)*SQUJ2          ! TRESCA
          CASE(2); YIELD=ROOT3*SQUJ2                  ! VON MISES
          CASE(3); SNPHI=SIN(FRICT)                   ! MOHR-COULOMB
            YIELD=SMEAN*SNPHI+SQUJ2*(COS(THETA)-SIN(THETA)*SNPHI/ROOT3)
            YIELD=YIELD/CEQS      ! Consistent form
          CASE(4); YIELD=(ALFA*3.*SMEAN+SQUJ2)/CEQS   ! DRUCKER-PRAGER
        END SELECT
        IF(NFL.EQ.0) RETURN
!         Evaluate derivatives of yield functions
        CALL DF_DSIG(SQUJ2,THETA,VARJ2,DEVIA,DFDS)
        END SUBROUTINE PL_FLOW
```

10.15 Subroutine DF_DSIG

This routine evaluates the derivatives of the yield function, $\partial f/\partial \sigma_{ij}$ (refer to Section 7.5.3 for details). In three dimensions, these are evaluated in the order: f_{11}, f_{22}, f_{33}, f_{12}, f_{23}, f_{31}; for plane strain problems the order is: f_{11}, f_{22}, f_{12}, f_{33} (the last term is omitted in plane stress). If the second deviatoric invariant J_2 is very small, the stress state is hydrostatic, which simplifies matters considerably.

Local variables

CONS1/2/3:	Coefficients $C_1/C_2/C_3$.
DFDS:	Vector $\partial f/\partial \sigma_{ij}$.
VECA1:	Vector $\partial I_1/\partial \sigma_{ij}$.
VECA2:	Vector $\partial \sqrt{J_2}/\partial \sigma_{ij}$.
VECA3:	Vector $\partial J_3/\partial \sigma_{ij}$.

```
        SUBROUTINE DF_DSIG(SQUJ2,THETA,VARJ2,DEVIA,DFDS)
        USE FIXED_VALUES
        REAL DEVIA(3,3),VECA1(6),VECA2(6),VECA3(6),DFDS(NSIG)
!                       Evaluate C1 and DI1-Dsigma
```

```
      CONS1=0.; IF(NCRIT.EQ.3) CONS1=SIN(FRICT)/3.
      IF(NCRIT.EQ.4) CONS1=ALFA
      VECA1(1:NDIM)=1.; VECA1(NDIM+1:NSIG)=0.
      IF(NSIG.EQ.4) VECA1(4)=1.                    ! Plane strain case
      IF(SQUJ2.LT.1.E-6) THEN                      ! Hydrostatic state
        DFDS(1:NSIG)=CONS1/CEQS*VECA1(1:NSIG)
        RETURN
      ENDIF
!                     Evaluate DJ2-Dsigma and DJ3-Dsigma
      TANTH=TAN(THETA); TANT3=TAN(3.*THETA); SINTH=SIN(THETA)
      COSTH=COS(THETA); COST3=COS(3.*THETA)
      L=0; DO 10 I=1,NDIM; DO 10 J=I,NDIM; L=L+1; M=INX(L)
      VECA2(M)=DEVIA(I,J)/(2.*SQUJ2)          !
      VECA3(M)=0.; DO 10 K=1,NDIM             !
10    VECA3(M)=VECA3(M)+DEVIA(I,K)*DEVIA(K,J)-VARJ2*DLT(I,J)*2./3.
      IF(NSIG.NE.4) GOTO 15
      VECA2(4)=DEVIA(3,3)/(2.*SQUJ2)    ! For plane strain
      VECA3(4)=DEVIA(3,3)*DEVIA(3,3)-VARJ2*2./3.
!              Find C2 and C3 for four yield criteria
15    CONS3=0.0; ABTHE=ABS(THETA*57.29577951308)
      SELECT CASE(NCRIT)
      CASE(1); IF(ABTHE.LT.29.) GO TO 20             ! TRESCA
        CONS2=ROOT3; GO TO 40
20    CONS2=2.*(COSTH+SINTH*TANT3); CONS3=ROOT3*SINTH/(VARJ2*COST3)
      CASE(2); CONS2=ROOT3                           ! VON MISES
      CASE(3); IF(ABTHE.LT.29.) GO TO 30             ! MOHR-COULOMB
        CONS2=0.5*ROOT3*(1.-SIGN(1.,THETA)*CONS1); GO TO 40
30    CONS2=COSTH*(1.+TANTH*TANT3+CONS1*(TANT3-TANTH)*ROOT3)
        CONS3=(ROOT3*SINTH+3.*CONS1*COSTH)/(2.*VARJ2*COST3)
      CASE(4); CONS2=1.                              ! DRUCKER-PRAGER
      END SELECT
!         Evaluate Df-Dsigma
40    IF(NCRIT.GT.2) THEN
        CONS1=CONS1/CEQS; CONS2=CONS2/CEQS; CONS3=CONS3/CEQS
      ENDIF
      DO 50 I=1,NSIG    !
50    DFDS(I)=CONS1*VECA1(I)+CONS2*VECA2(I)+CONS3*VECA3(I)
      END SUBROUTINE DF_DSIG
```

10.16 Subroutine P_M_ITER

This routine solves the nonlinear system equations, expressed in terms of the plastic multiplier representation of Section 9.6.2, using the Newton–Raphson iterative scheme of Section 9.7.1. For computational purposes, some further manipulation of the results of Section 9.6.2 is necessary. Thus, substituting Eq. (9.42) into (9.40), we obtain

$$[A^\lambda] = [I] - [\nabla f_\Psi][d^f] - [\nabla f_\Psi][E][d^f] \qquad (10.4)$$

Using Eqs. (9.28), (9.30), and (7.68), we can simplify the above equation by using the substitution

$$[\nabla f_\Psi][d^f] = 1 - H^* h_{\varepsilon,\lambda} / \Psi \qquad (10.5)$$

where $h_{\varepsilon,\lambda}$ is obtained from Eq. (7.50).

Local variables

A:	Matrix $[A^\lambda]$.
CBACK:	$\dfrac{\overline{H}}{\sigma_y}(1 - m)\dot{\bar{\varepsilon}}^{\mathrm{p}}$ (from Section 7.7).
CLAMD:	$1 - H^* h_{\varepsilon,\lambda} / \Psi$.
COEF:	Working array.
DF:	Vector $\{d^f\}$ (Eq. 9.28).
DFDS:	Vector of $\partial f / \partial \sigma_{ij}$.
DLAMD:	Vector of $\{\dot{\lambda}\}$.
DPLAS:	Contains $\dot{\bar{\varepsilon}}^p$ (from Section 7.5.2).
EQUP2:	$\dfrac{\partial f}{\partial \sigma_{ij}} \dfrac{\partial f}{\partial \sigma_{ij}}$.
ESTRS:	Array of effective stresses, $\sigma_{ij} - \rho_{ij}$.
FINCR:	Loading factor φ.
FMULT:	Loading multiplier φ^{n-1} for the nth increment.
FSCAL:	Scaling factor for output of specified tractions.
HEPSI:	$h_{\varepsilon,\lambda}$ (Eq. 7.50).
NLAMD:	Number of current yielded nodes.
NNOD:	Vector of yielded node numbers (0 = elastic state).
PSI:	Ψ (Eq. 7.68).
RERR:	Relative error ($= \sqrt{R^T R}/\sigma_y$).
RNORM:	The square norm ($R^T R$) of the residue (Eq. 9.46).
YE:	$\{\dot{y}^e\}$ (Eq. 9.35).

```
      SUBROUTINE P_M_ITER(NODE,TOL,FSCAL,WOVER)
      USE VARY_ARRAYS,YE=>YSIG,DLAMD=>Y,A=>CD; USE FIXED_VALUES
      DOUBLE PRECISION SS
!              FORM INITIAL COEFFICIENT MATRICES
      CALL MATRICES(NBF,NTF,NTSF,COEF,COEF(1,NTP+1),DLAMD,YE)
      IF(NOUT1.NE.0)WRITE(7,'(/" RESULTS BEFORE INCREMENT LOADING:")')
      IF(NOUT1.NE.0) CALL OUTPUT(NODE,NOUT1,1,FSCAL)
      FMULT=1.    ! INITIAL VALUE OF INEQUAL INCREMENT FACTOR
      DO 90 INCRS=1,NINCS       ! Loop for load increment
      FSCAL=FSCAL+WOVER*FMULT   !
!              Scale trial stress to yield surface
      CALL SIG_SCALE(ESTRS,EPSTN,X,COEF,TOL,FMULT,YE,NNOD,NLAMD)
```

```
         ALLOCATE (A(NLAMD,NLAMD+1)); IF(NLAMD.EQ.0) GOTO 85
         DLAMD=0.
         DO 10 IP=1,NTP          ! Find Df-Dsigma
         CALL PL_FLOW(ESTRS(1,IP),YIELD,DFDS(1,IP),1)
         DO 10 I=1,NSIG  ! Update stresses for elastic part
10       ESTRS(I,IP)=ESTRS(I,IP)+YE(NSIG*(IP-1)+I)
         DO 70 ITER=1,IABS(MITER)    ! Loop for iteration
!                    Form [A] AND [Y]->[A(I,NTP+1)]
         REWIND(56)    !
         DO 60 IP=1,NTP; IL=NNOD(IP)
         DO 20 I=1,NSIG; IF(IL.EQ.0) READ(56)
20       IF(IL.NE.0) READ(56)(COEF(I,J),J=1,NTSF)    ! Matrix [E]
         IF(IL.EQ.0) GOTO 60
         IO=NSIG*(IP-1); CLAMD=0.
         DO I=1,NLAMD+1; A(IL,I)=0.; ENDDO
         CALL DF_MATRX(DFDS(1,IP),PSI,HEPSI)    ! Find ipsi
         CLAMD=1.-HARDS*HEPSI/PSI
         DO 30 I=1,NSIG         !
30       A(IL,NLAMD+1)=A(IL,NLAMD+1)+RE(I)*DFDS(I,IP)*YE(IO+I)/PSI
         DO 50 JP=1,NTP; JL=NNOD(JP); IF(JL.EQ.0) GOTO 50
!              Find [De] and Df-Dsigma
         CALL DF_MATRX(DFDS(1,JP),PSIB,HEPSI)
!          Find last term of equation
         JO=NSIG*(JP-1)
         DO 40 I=1,NSIG; DO 40 J=1,NSIG
40       A(IL,JL)=A(IL,JL)-RE(I)*DFDS(I,IP)/PSI*COEF(I,JO+J)*DF(J)
50       CONTINUE
!      Form matrix [A]
         A(IL,IL)=1.-CLAMD+A(IL,IL)
60       CONTINUE
!                        CHECK CONVERGENCE
         RNORM=0.
         DO IP=1,NLAMD; SS=DBLE(A(IP,NLAMD+1))   ! Find {Yf}
           DO JP=1,NLAMD
             SS=SS-DBLE(A(IP,JP))*DBLE(DLAMD(JP))    ! Form {R}
           ENDDO
           A(IP,NLAMD+1)=SS               !
           RNORM=RNORM+SS*SS              ! Calculate {R}.{R}
         ENDDO
         RERR=SQRT(RNORM)/UNIAX      ! Calculate relative error
         WRITE(*,95)RERR,NLAMD,ITER,INCRS
         WRITE(7,95)RERR,NLAMD,ITER,INCRS
         IF(RERR.LT.TOLER) GOTO 75       ! Check convergence
!             Solving system equations
         CALL INVSOLVR(NLAMD,NLAMD+1,A,NLAMD,1)
```

```
!                    UPDATE VARIABLES
      DO IP=1,NLAMD            ! Update plastic multipliers
        DLAMD(IP)=DLAMD(IP)+A(IP,NLAMD+1)    !
      ENDDO
      CALL UPDATEV(NLAMD,A(1,NLAMD+1))    ! Update {X}, {u} and {sigma}
      DO IP=1,NTP         ! Update Df-Dsigma
        CALL PL_FLOW(ESTRS(1,IP),YIELD,DFDS(1,IP),1)
      ENDDO
70    CONTINUE   ! END OF ITERATION LOOP
!            Update equivalent plastic strain and stresses
75    DO 80 IP=1,NTP; EQUP2=0.
      IF(NNOD(IP).EQ.0)GOTO 80                   ! IP is in elastic state
      IF(DLAMD(NNOD(IP)).LE.0.)GOTO 80        ! Unacceptable case
      DO I=1,NSIG; EQUP2=EQUP2+RE(I)*DFDS(I,IP)*DFDS(I,IP); ENDDO
      DPLAS=CEQP*SQRT(EQUP2)*DLAMD(NNOD(IP))   !
      EPSTN(IP)=EPSTN(IP)+DPLAS               ! Update equivalent plastic strain
      CBACK=DPLAS*HARDK/(UNIAX+HARDI*EPSTN(IP))   !
      BACKS(:,IP)=BACKS(:,IP)+CBACK*ESTRS(:,IP)        ! Update back stress
      ESTRS(:,IP)=ESTRS(:,IP)-CBACK*ESTRS(:,IP)        ! Update
80    CONTINUE
!                    Output results
85    IF(INCRS.LT.NINCS) CALL OUTPUT(NODE,NOUT1,INCRS,FSCAL)
      DEALLOCATE (A)
!          Form loading factor for next increment
90    FMULT=FMULT*FINCR     !
95    FORMAT(//5X,'RERR ==',1PE14.6,', NYIELD ==', I4,                    &
      &        ',  ITER ==',I3,',    INCREM == ',I3)
      END SUBROUTINE P_M_ITER
```

10.17 Subroutine MATRICES

This routine forms the initial stress coefficient matrices for the system equations
formulated in terms of plastic multipliers. These are $[A^c]$, $[E]$, and $[E_u^I]$ defined in
Eqs. (9.34), (9.37), and (9.45), respectively. These matrices are stored on Channels
54 and 56.

Local variables

AC:	Array containing $[A^c]$ and $[E^b]$.
ASIG:	Vector containing $[A^\sigma]$ and $[A^b]^{-1}$.
AU:	Vector containing $[A^u]$.
ESIG:	Vector containing $[E^\sigma]$, $[A^c]$, and $[E]$.
EU:	Vector containing $[E^u]$ and $[E_u^I]$.
NTFB:	Number of interior displacement degrees of freedom.

```
SUBROUTINE MATRICES(NBF,NTF,NTSF,Asig,Esig,Au,Eu)
REAL Asig(NBF),Esig(NTSF),Au(NBF),Eu(NTSF),Ac(NBF,NTSF)
DOUBLE PRECISION SS1,SS2
NTFB=NTF-NBF; REWIND(51); REWIND(56)
DO I=1,NBF; READ(51)(Ac(I,J),J=1,NTSF); ENDDO      ! [Eb]
DO 20 I=1,NBF; READ(56)(Asig(J),J=1,NBF)   !   [Ab]inverse
DO 10 J=1,NTSF; SS1=0.0D0
DO K=1,NBF; SS1=SS1+DBLE(Asig(K))*DBLE(Ac(K,J)); ENDDO
10      Esig(J)=SS1                        ! [Ab]inverse[Eb]
20      WRITE(54)(Esig(J),J=1,NTSF)        ! [Ac]=[Ab]inverse[Eb]
REWIND(54); DO I=1,NBF; READ(54)(Ac(I,J),J=1,NTSF); ENDDO
!                 FORM [E]->(56),          [Eu]-->(54)
REWIND(50); REWIND(52); REWIND(53); REWIND(56)
DO 50 I=1,NTSF
READ(52)(Asig(J),J=1,NBF)      ! Retrieve matrix [A-sigma]
READ(53)(Esig(J),J=1,NTSF)     ! Retrieve matrix [E-sigma]
IF(I.LE.NTFB) READ(51)(Eu(J),J=1,NTSF)  ! Read  [Eu]
IF(I.LE.NTFB) READ(50)(Au(J),J=1,NBF)   ! READ  [Au]
DO 40 J=1,NTSF; SS1=DBLE(Eu(J)); SS2=DBLE(Esig(J))
DO 30 K=1,NBF
IF(I.LE.NTFB) SS1=SS1+DBLE(Au(K))*DBLE(Ac(K,J))  ! [Au][Ac]
30      SS2=SS2+DBLE(Asig(K))*DBLE(Ac(K,J))       ! [Asigma][Ac]
        Eu(J)=SS1              ! [Eu]=[Eu]+[Au][Ac]->Eu
40      Esig(J)=SS2            ! [E]=[E-sigma]+[A-sigma][Ac]-->Esig
IF(I.LE.NTFB) WRITE(54)(Eu(J),J=1,NTSF)   ! [Eu]-->(54)
50      WRITE(56)(Esig(J),J=1,NTSF)        ! [E]-->(56)
DO I=50,53; CLOSE(I,STATUS='DELETE'); ENDDO
END SUBROUTINE MATRICES
```

10.18 Subroutine SIG_SCALE

This routine scales the stress predictor $\{\sigma^t\}$ to the yield surface (refer to Section 9.7.2). Yielding nodes are also identified. Quantities stored on Channel 55 are the values for the first increment. Thus, the incremental values in the nth increment can be obtained by multiplying them by φ^{n-1} (see Section 9.7.3).

Local variables

ESTRS: Array of the effective stresses, $\sigma_{ij} - \rho_{ij}$.
FGRAD: Working vector.
FMULT: Load multiplier φ^{n-1}.
FSCAL: Scaling factor for output of specified tractions.
NLAMD: Number of current yielded nodes.
NNOD: Vector of yield node numbers (0 = elastic state).
ROVER: Excess stress factor.
SIGMT: Vector containing stress predictor $\{\sigma^t\}$.

SIGY: Current yield stress.
TOL: Tolerance factor for scaling stress to yield.
YC: $\{\dot{y}^c\}$.
YE: $\{\dot{y}^e\}$.
YIELD: Equivalent effective stress for current stress predictor.
YILD0:   Equivalent effective stress at end of previous increment.

```
      SUBROUTINE SIG_SCALE(ESTRS,EPSTN,X,Yc,TOL,FMULT,YE,NNOD,NLAMD)
      USE FIXED_VALUES,SIGMT=>STRES
      DIMENSION ESTRS(NSIG,NTP),EPSTN(NTP),X(NTF),YE(NTSF),NNOD(NTP),   &
     &               Yc(NTF),FGRAD(6)
      REWIND(55); READ(55)Yc; READ(55)YE ! Get data from increment 1
      NLAMD=0              ! For accumulating yield nodes
      X=X+Yc*FMULT         ! Update unknowns by adding incremental value
      DO 90 IP=1,NTP; IO=NSIG*(IP-1)
      NNOD(IP)=0           ! Assume node IP in elastic state
      ROVER=1.             ! Assume whole current increment beyond yield
      DO I=1,NSIG
        SIGMT(I) =ESTRS(I,IP) +YE(IO+I) *FMULT          !
      ENDDO
      SIGY=UNIAX+HARDI*EPSTN(IP)  ! Current yield stress limit
      CALL PL_FLOW(ESTRS(1,IP),YILD0,FGRAD,0) ! Previous equiv. stress
      CALL PL_FLOW(SIGMT,YIELD,FGRAD,0)        ! Predicted equiv. stress
      IF(YIELD.LE.SIGY) THEN; ROVER=0.; GOTO 60; ENDIF ! Elastic loading
      IF(YILD0.LT.SIGY-TOL) GOTO 30            ! Previous state is elastic
      GOTO 50              ! Both previous and current state are plastic
!              Determine scale factor (alpha) to yield surface
30    CALL SIGCROSS(ESTRS(1,IP),SIGY,YILD0,YIELD,ROVER)
50    NLAMD=NLAMD+1        ! Accumulate the number of yielded nodes
      NNOD(IP)=NLAMD       ! Record the position of this yield node
!     Scale stresses to yield surface and set up excess load
60    DO 80 I=1,NSIG        !
      ESTRS(I,IP)=ESTRS(I,IP)+YE(IO+I)*FMULT*(1.-ROVER)  ! Scale stress
80    YE(IO+I)=YE(IO+I)*FMULT*ROVER ! set up excess loads
90    CONTINUE
      END SUBROUTINE SIG_SCALE
```

10.19 Subroutine SIGCROSS

This routine calculates the position of the intersection of the stress predictor with the yield surface using the iterative method described in Section 9.7.2.

Local variables
ALFAC: α.
ALFAP: α^0.

FGRAD: $\partial \bar{f}/\partial \sigma$.
NITER: Maximum number of iterations.
RESFC: $\bar{f}(\sigma_n + \alpha \Delta \sigma) - \sigma_Y$.
ROVER: $1 - \alpha$.
SIGMA: $\sigma_n + \alpha \Delta \sigma$.
SIGMN: σ_n.
SIGMT: σ^t.
SIGY: σ_y.
TOLE: Iteration tolerance.
YIELD: $\bar{f}(\sigma^t)$.
YILD0: $\bar{f}(\sigma_n)$.

```
           SUBROUTINE SIGCROSS(SIGMN,SIGY,YILD0,YIELD,ROVER)
           USE FIXED_VALUES,SIGMT=>STRES
           REAL SIGMN(NSIG),SIGMA(6),FGRAD(6)
           TOLE=1.E-5; NITER=10
           ALFAP=(SIGY-YILD0)/(YIELD-YILD0); ALFAC=ALFAP    !
           RESF=ABS(YIELD-SIGY)
           DO 50 ITER=1,NITER
           DO I=1,NSIG; SIGMA(I)=SIGMN(I)+ALFAC*(SIGMT(I)-SIGMN(I)); ENDDO
!              Compute current equivalent stress and Df-Dsigma
           CALL PL_FLOW(SIGMA,YIELD,FGRAD,1)
           RESFC=YIELD-SIGY
           IF(ABS(RESFC).LT.TOLE) GOTO 60       ! On yield surface
!                  Update alpha
           IF(ABS(RESFC).LT.RESF) THEN
            ALFAP=ALFAC; RESF=ABS(RESFC)
           ENDIF
!                                          Compute increment in alpha
           EQ=0.
           DO I=1,NSIG
            EQ=EQ+RE(I)*FGRAD(I)*(SIGMT(I)-SIGMN(I))
           ENDDO
!                                          Compute current value of alpha
50         ALFAC=ALFAC-RESFC/EQ
           ALFAC=ALFAP           !          Update alpha
60         ROVER=1.-ALFAC        !          Overload factor
           END SUBROUTINE SIGCROSS
```

10.20 Subroutine DF_MATRX

This routine calculates the matrix $\{d^f\}$ from Eq. (9.28), the quantity $h_{\varepsilon,\lambda}$ from Eq. (7.50), and the parameter Ψ from Eq. (7.68). To compute the first of these quantities, we take advantage of the fact that the fourth-order constitutive tensor D^e_{ijkl} can, for elastic isotropic materials, be expressed in terms of a second-order tensor. Thus, we obtain

$$d_{ij}^f = D_{ijkl}^e \frac{\partial f}{\partial \sigma_{kl}}$$

$$= \lambda \delta_{ij} \frac{\partial f}{\partial \sigma_{kk}} + 2G \frac{\partial f}{\partial \sigma_{ij}}$$
(10.6)

The terms of the vector $\{d^f\}$ are ordered in the usual sequence. Hence, in three dimensions, the order is: $d_{11}, d_{22}, d_{33}, d_{12}, d_{23}, d_{31}$; for plane strain problems the order is: $d_{11}, d_{22}, d_{12}, d_{33}$ (the last term is omitted in plane stress). In this subroutine, the above equation is normalized by dividing throughout by $2G$.

Local variables

AVKK: $\partial f / \partial \sigma_{kk}$.

CLAME: The Lame parameter λ.

DFDS: Vector of $\partial f / \partial \sigma_{ij}$.

EQUP2: $\dfrac{\partial f}{\partial \sigma_{ij}} \dfrac{\partial f}{\partial \sigma_{ij}}$.

HEPSI: $h_{\varepsilon,\lambda}$ (Eq. 7.50).

PSI: Ψ (Eq. 7.68).

```
      SUBROUTINE DF_MATRX(DFDS,PSI,HEPSI)
      USE FIXED_VALUES; REAL DFDS(NSIG)
      AVKK=SUM(DFDS(1:NDIM))   ! For 3D and plane stress problems
!          For plane strain problems, additional term needed
      IF(NSIG.EQ.4) AVKK=AVKK+DFDS(4)
      L=0; DO 20 I=1,NDIM; DO 20 J=I,NDIM; L=L+1
20    DF(INX(L))=CLAME*DLT(I,J)*AVKK+DFDS(INX(L))  !
      IF(NSIG.EQ.4) DF(4)=CLAME*AVKK+DFDS(4)   ! For plane strain
      PSI=0.; EQUP2=0.
      DO 30 I=1,NSIG
      EQUP2=EQUP2+RE(I)*DFDS(I)*DFDS(I)
30    PSI=PSI+RE(I)*DFDS(I)*DF(I)          ! First term in equation
      HEPSI=CEQP*SQRT(EQUP2)         !
      PSI=PSI+HARDS*HEPSI            !
      END SUBROUTINE DF_MATRX
```

10.21 Subroutine UPDATEV

This routine updates values of boundary unknowns, interior displacements, and stresses during iteration. The boundary unknowns and interior displacements are calculated.

Local variables

COEF: Working array.

DF: Vector $\{d^f\}$.

DLAMD: Vector $\{\dot{\lambda}\}$.

NLAMD: Number of current yielded nodes.
NNOD: Vector of yielded node numbers.
X: Vector of boundary unknowns, interior displacements.

```
       SUBROUTINE UPDATEV(NLAMD,DLAMD)
       USE VARY_ARRAYS; USE FIXED_VALUES
       DOUBLE PRECISION SS,S(6); REAL DLAMD(NLAMD)
!       Compute and update boundary unknowns and interior displacements
       REWIND(54)       ! [Ac] and [Eu] were stored in subroutine MATRICES
       DO 30 I=1,NTF; SS=DBLE(X(I))
       READ(54)(COEF(1,J),J=1,NTSF)    ! [Ac] or [Eu]
       DO 20 JP=1,NTP; J0=NSIG*(JP-1)
       IF(NNOD(JP).EQ.0) GOTO 20        ! Node JP has not yielded
       CALL DF_MATRX(DFDS(1,JP),PSI,HEPSI)     ! Find [df]
       DO 10 J=1,NSIG       !
10     SS=SS+DBLE(COEF(1,J0+J))*DBLE(DF(J))*DBLE(DLAMD(NNOD(JP)))
20     CONTINUE
30     X(I)=SS         ! Update boundary unknowns or interior displacements
!                            Compute and update stresses
       REWIND(56)   ![E] was stored on channel 56 in subroutine MATRICES
       DO 70 IP=1,NTP; DO 40 I=1,NSIG; S(I)=DBLE(ESTRS(I,IP))
40     READ(56)(COEF(I,J),J=1,NTSF)     ! Retrieve [E]
       DO 60 JP=1,NTP; J0=NSIG*(JP-1)
       IF(NNOD(JP).EQ.0) GOTO 60        ! Skip nodes which have not yielded
       CALL DF_MATRX(DFDS(1,JP),PSI,HEPSI)     ! Find [df]
       DO 50 I=1,NSIG; DO 50 J=1,NSIG
!
50     S(I)=S(I)+DBLE(COEF(I,J0+J))*DBLE(DF(J))*DBLE(DLAMD(NNOD(JP)))
60     CONTINUE
       ESTRS(1:NSIG,IP)=S(1:NSIG)          ! Update stresses
70     CONTINUE
       END SUBROUTINE UPDATEV
```

10.22 Closure

It is evident that understanding this complex code, which implements the nonlinear numerical algorithm described in the previous chapter, is not an easy proposition. We have attempted to provide a reasonably comprehensive, and structured, guide to the code in this chapter, but full understanding will also, inevitably, require careful study of most of the book. The code and explanatory matter for specific subroutines will be helpful to those interested in developing an understanding of how aspects of the theoretical and numerical formulations of the previous chapters can be translated into concrete form. Before running the program in earnest, we recommend a review of the description of the input data in Appendix H, and some of the example problems described in the following chapter.

CHAPTER ELEVEN

Nonlinear Applications

11.1 Introduction

In this chapter, we present some applications of the computer code described earlier. Some benchmark examples are used to verify its validity. More complex practical problems, involving over 5,000 degrees of freedom, are used to demonstrate the practical utility of the code.

11.2 A Cube Subjected to Uniaxial Tension

In this example, we analyze the response of a cube (Fig. 11.1) subjected to uniaxial tensile displacement. This benchmark example is intended to illustrate isotropic, kinematic, and mixed hardening (including perfect plasticity and softening phenomena) under uniaxial loading. The Von Mises criterion is assumed to apply.

The cube was discretized using 24 linear boundary elements (4 per face), defined by 26 nodes, and 8 linear volume cells. The latter required specification of one more node (at the cube center), making a total of 27 nodes. The "roller" boundary condition (zero normal displacement, zero shear stress) was imposed on the three planes, $x = 0$, $y = 0$, and $z = 0$. The cube was then subjected to monotonically increasing displacement to the upper ($z = 10$) surface. The input for the isotropic hardening case ($H' = 0.1, \overline{H} = 0.05, m = 1$) is reproduced below. (These data are also contained in the file CUBE3D.DAT; the output data may be inspected in its entirety in the file CUBE3D.OUT.)

```
A CUBE UNDER EXTENSION
     6     4    26    27    24     8     0     2     0    -6
     -.00001        .0        .0        .0
     1      10.0       10.0       10.0
     2       5.0       10.0       10.0
     3        .0       10.0       10.0
     4      10.0        5.0       10.0
```

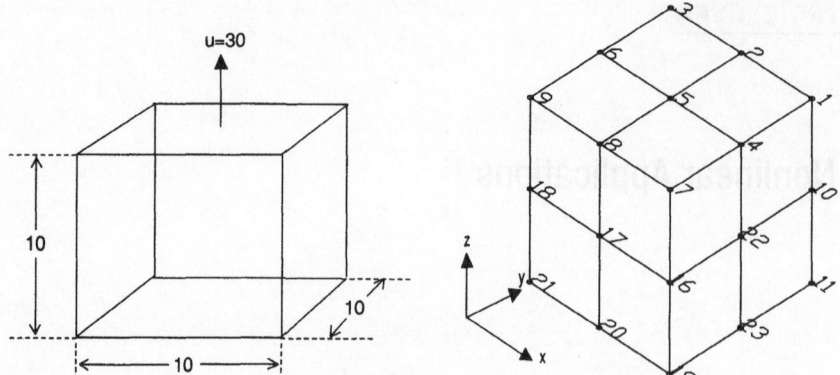

Figure 11.1: A cube subjected to uniaxial tension.

5	5.0	5.0	10.0
6	.0	5.0	10.0
7	10.0	.0	10.0
8	5.0	.0	10.0
9	.0	.0	10.0
10	10.0	10.0	5.0
11	10.0	10.0	.0
12	5.0	10.0	5.0
13	5.0	10.0	.0
14	.0	10.0	5.0
15	.0	10.0	.0
16	10.0	.0	5.0
17	5.0	.0	5.0
18	.0	.0	5.0
19	10.0	.0	.0
20	5.0	.0	.0
21	.0	.0	.0
22	10.0	5.0	5.0
23	10.0	5.0	.0
24	.0	5.0	5.0
25	.0	5.0	.0
26	5.0	5.0	.0
27	5.0	5.0	5.0

23		
1	1	1
2	1	1
3	1	101
4	1	1
5	1	1
6	1	101
7	1	11
8	1	11

9	1	111
11	2	1
13	2	1
14	2	100
15	2	101
16	2	10
17	2	10
18	2	110
19	2	11
20	2	11
21	2	111
23	2	1
24	2	100
25	2	101
26	2	1

0

1	1	2	5	4	0	110
2	2	3	6	5	0	110
3	4	5	8	7	0	110
4	5	6	9	8	0	110
5	1	10	12	2	0	111
6	10	11	13	12	0	111
7	2	12	14	3	0	111
8	12	13	15	14	0	111
9	7	8	17	16	0	101
10	8	9	18	17	0	101
11	16	17	20	19	0	101
12	17	18	21	20	0	101
13	1	4	22	10	0	111
14	4	7	16	22	0	111
15	10	22	23	11	0	111
16	22	16	19	23	0	111
17	3	14	24	6	0	11
18	14	15	25	24	0	11
19	6	24	18	9	0	11
20	24	25	21	18	0	11
21	15	13	26	25	0	110
22	13	11	23	26	0	110
23	25	26	20	21	0	110
24	26	23	19	20	0	110

1	.0	.0	30.0
2	.0	.0	.0

.0   1.0   0.3

1	11	13	26	23	10	12	27	22
2	13	15	25	26	12	14	24	27

Table 11.1. Stresses, strains, and displacements at cube top face

	σ_{zz}	ε_{zz}^p	u_z	u_x
Numerical	1.0000	2.0000	30.000	−6.5001
Analytical	1.0	2.0	30.0	−6.5

```
3  23  26  20   19    22  27  17  16
4  26  25  21   20    27  24  18  17
5  10  12  27   22     1   2   5   4
6  12  14  24   27     2   3   6   5
7  22  27  17   16     4   5   8   7
8  27  24  18   17     5   6   9   8
2  0.8  20.  0.1  0.05   1.0
6  1.0  10  0.001 3303  1101
```

The principal results (for stresses, plastic strains, and displacements), when the imposed displacement is 30 units, are given in Table 11.1, for the center of the cube's top face.

Figure 11.2 depicts the vertical stress developed at the cube's top face in response to the imposed displacement and, also, the computed equivalent stress versus the equivalent plastic strain. These results are essentially exact.

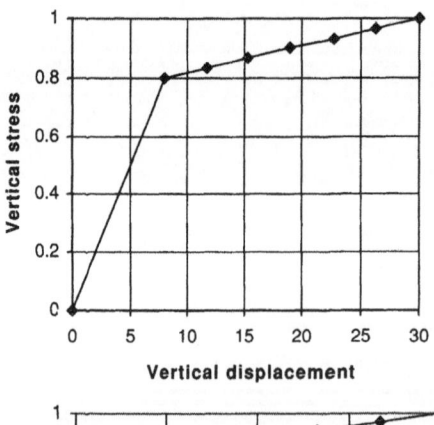

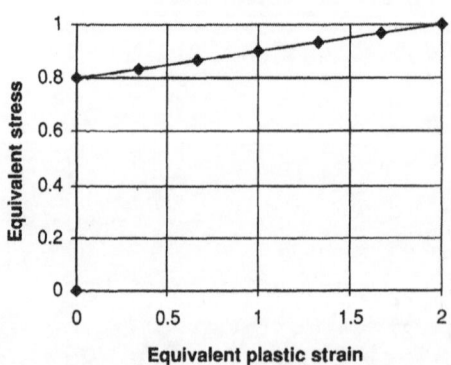

Figure 11.2: The isotropic hardening response.

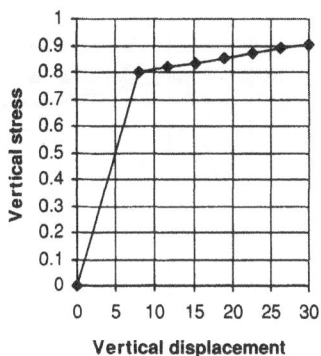

Figure 11.3: The kinematic hardening response.

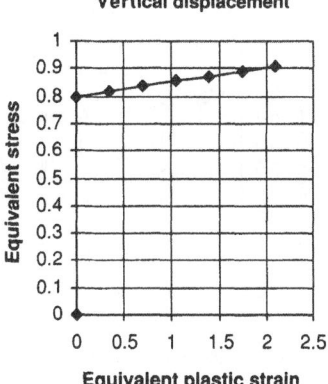

For a kinematic hardening material, the input data remains unchanged, with the exception that the proportionality parameter (m) now takes the value zero rather than unity. The corresponding results for this case are depicted in Fig. 11.3.

Finally, three cases of mixed hardening were analyzed, using the parameters listed in Table 11.2. Of course, the remaining input data are unchanged from that employed in the previous case. Figure 11.4 depicts the results obtained in these three cases. As expected, and as for the isotropic and kinematic hardening cases, the results are essentially exact in all respects. This three-dimensional benchmark example demonstrates, at least for a theoretically homogenous deformation mode, that the program code handles hardening and softening materials properly.

11.3 A Thick-Walled Cylinder Subjected to Internal Pressure

In this example, a thick-walled cylinder (Fig. 11.5) is subjected to internal pressure under plane strain conditions. This problem has already been analyzed, assuming

Table 11.2. Parameters for the mixed hardening case

	m	H'	$\overline{H}$	H^*
Case 1	0.5	0.1	0.05	0.075
Case 2	2/3	−0.1	0.05	−0.05
Case 3	1/3	−0.1	0.05	0

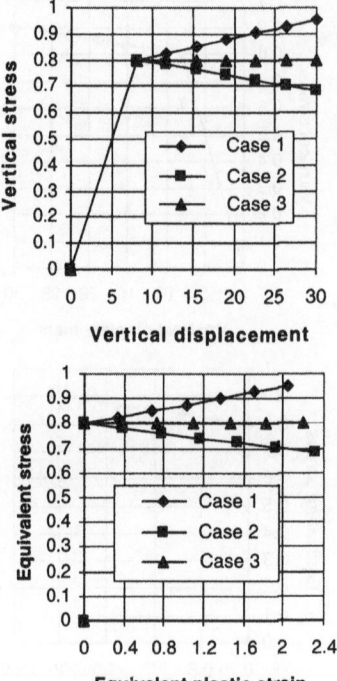

Figure 11.4: The mixed hardening response.

an elastic response, in Chapter 6. A quadrant of the cylinder is discretized using 12 quadratic boundary elements and 36 cells, defined by a total of 133 nodes. Of these, 26 are boundary nodes and 107 are interior nodes. The Tresca yield criterion with perfect plasticity is assumed here. (The input data and the output data are contained in the files CYLP2D.DAT and CYLP2D.OUT, respectively.)

The radial displacement response to the applied pressure is shown in Fig. 11.6. The radial and circumferential (hoop) stress distributions for a specific internal pressure ($p = 20$ units) are plotted in Fig. 11.7. On these scales, the theoretical (Lubliner, 1990) and numerical results are largely indistinguishable, except that some discrepancies (1–2%) are apparent in the hoop stresses. These are confined to the higher load levels and in the interior, whereas stresses at the outer boundary are

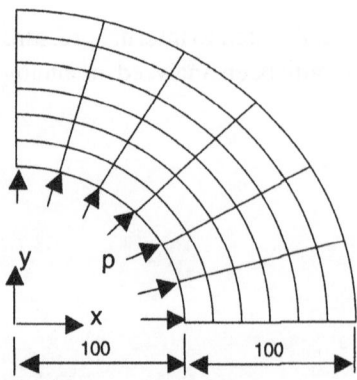

Figure 11.5: An internally pressurized thick cylinder.

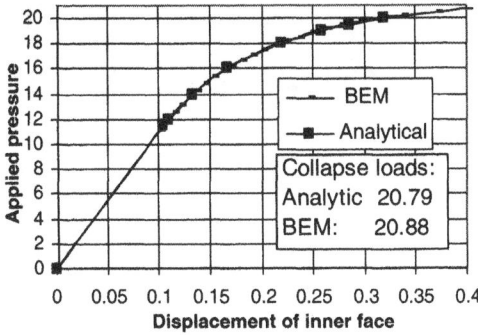

Figure 11.6: Radial displacement of the inner surface.

well captured. The reasons for these discrepancies have not been fully investigated but may be due to direct or consequential effects of the geometrical discretization.

11.4 A Rigid Punch under Plane Strain

This problem was the subject of a benchmark finite element study carried out on behalf of NAFEMS – the UK association concerned with finite element methods and standards – and reported by Linkens (1993). A rigid punch is impressed into a

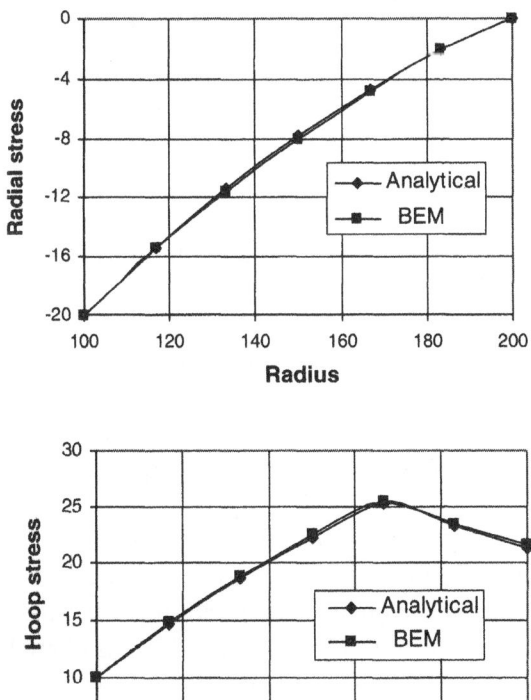

Figure 11.7: Radial and hoop stresses along radius (at $p = 20$).

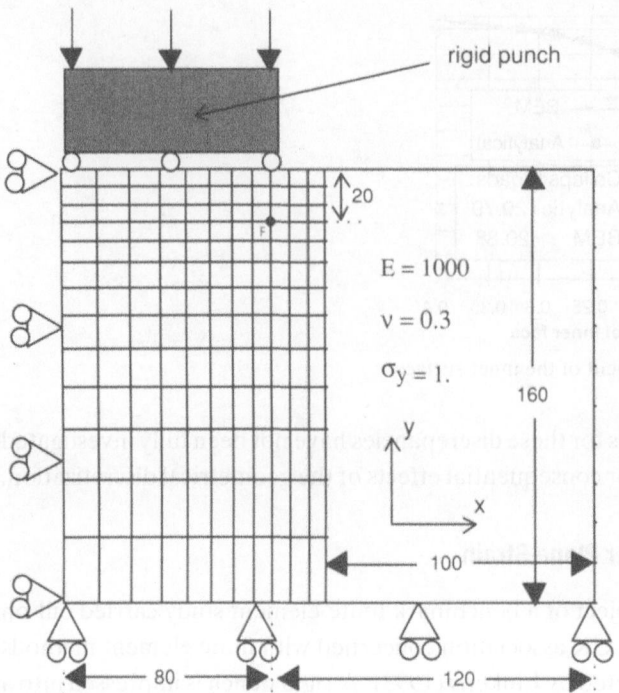

Figure 11.8: Rigid punch under plane strain.

finite continuum under plane strain conditions (Fig. 11.8). This results in a complex stress distribution and, in particular, stress concentrations beneath the edges of the punch. Not surprisingly, this problem defies analytical solution.

For simplicity, if we assume that the contact is frictionless, the boundary conditions can be simulated by monotonically increasing the vertical displacements at the contact nodes. Taking symmetry into account, the mesh discretization required specification of 528 nodes for the 60 quadratic boundary elements and 143 cells. Of these nodes, 121 were defined on the boundary; one of these is an additional node at the traction discontinuity at the edge of the punch. Calculations were carried out, assuming the Von Mises criterion, for two cases: perfect plasticity ($H' = 0$) and isotropic hardening ($H' = 0.111 E$). For the perfect plasticity case only, the input data and output data are contained in the files PUNCH.DAT and PUNCH.OUT.

The load–displacement plot for the perfect plasticity case is shown in Fig. 11.9a (we follow the NAFEMS publication by plotting the load on one side of the symmetry plane only). The vertical stress developed at the specific point F, which is located just below the edge of the punch, at the coordinate (80,20), is plotted in Fig. 11.9b. The finite element method (FEM) results shown in these figures are the benchmark results alluded to earlier. However, in the FEM study, the stresses at F were obtained by extrapolation from the Gauss points of the element to the upper left of the point F, a procedure that is bound to incur some error as the stress gradients are high here. Two sets of FEM results are shown in some of

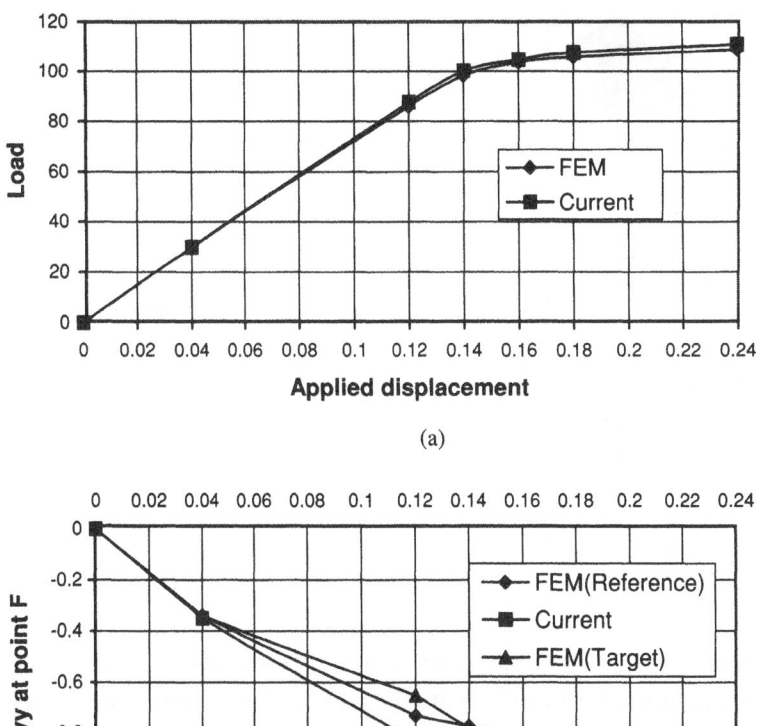

(a)

(b)

Figure 11.9: Numerical results for perfect plasticity model.

these figures: The set of results referred to as "target" were obtained using a finite element discretization scheme very similar to that employed using the BEM; that denoted by "reference" is believed to have emerged from a convergence study, which used much finer discretization schemes.

Comparing the results obtained by these methods, we observe that Fig. 11.9a demonstrates the very good agreement between the FEM (reference) solution and our current results. Determining the vertical stress at the point F is a much more demanding test (Fig. 11.9b) and the target results are unconvincing, since two inflection points may be observed in the data. The BEM results are certainly much closer to the reference solution and, despite some reservations about the accuracy of the latter, they provide encouraging evidence of the method's stability and reliability. The question of whether the BEM solution would converge to that of the reference solution remains open. The stress (and strain) concentrations that are induced at the punch edges are depicted in Fig. 11.10, in terms of contours

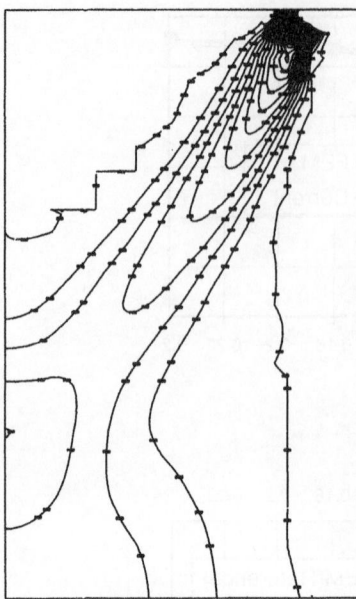

Figure 11.10: Equivalent plastic strain contours.

of equivalent plastic strain. These show, as expected, that a well-defined curved surface of intense shearing develops beneath the punch: when yielding takes place along the entire surface, failure is imminent.

For completeness, we also plot results obtained (in Fig. 11.11) by assuming the hardening plasticity material model and compare these with the FEM reference benchmark solutions. Again, the agreement is very good indeed and is certainly much better than that obtained using a comparable FEM discretization scheme.

11.5 A Flexible Square Footing

This three-dimensional example pertains to the behavior of a vertically loaded square footing (with dimension $B = 1$), up to collapse, founded on the surface of a half-space. It is conventional practice, in foundation engineering, to express the collapse load in terms of the so-called undrained shear strength (C_u) of the soil. From the definition of equivalent stress σ_y (uniaxial yield limit), which characterizes the von Mises yield criterion, it can be easily shown that $\sigma_y = 2C_u$. In this example, the undrained shear strength is taken to be unity, and the elastic material parameters (Young's modulus E, Poisson's ratio ν) were taken to be 1,000 and 0.3, respectively. In the BEM model, the far-field ground surface was simply discretized using progressively larger boundary elements and, at each node shared by the footing and ground surface, two coincident nodes were used to model the traction discontinuity. Uniform vertical pressure was applied to the footing and the quadrantal symmetry was exploited. The surface was discretized into just 57 quadratic boundary elements (including 9 elements over the footing), and 198 cells were defined in the expected yield zone. This required specification of

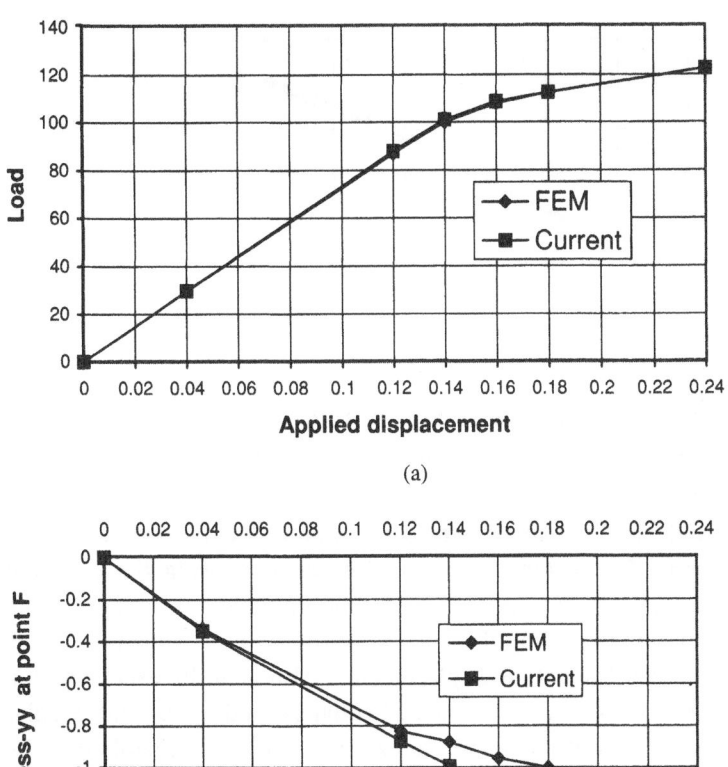

(a)

(b)

Figure 11.11: Numerical results for a hardening material.

nearly 1,200 nodes, of which 200 were defined on the surface. Part of the near-field discretization scheme is depicted in Fig. 11.12, both in plan and in elevation. (The complete data set is contained in the file FOOT.DAT and the output data may be found in FOOT.OUT.)

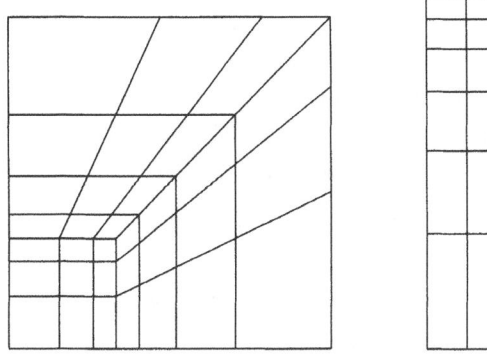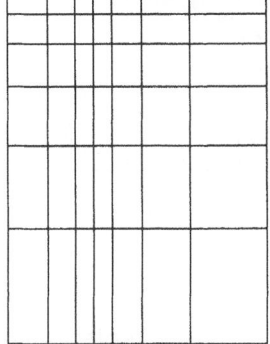

Figure 11.12: Plan and elevation of boundary elements and internal cells.

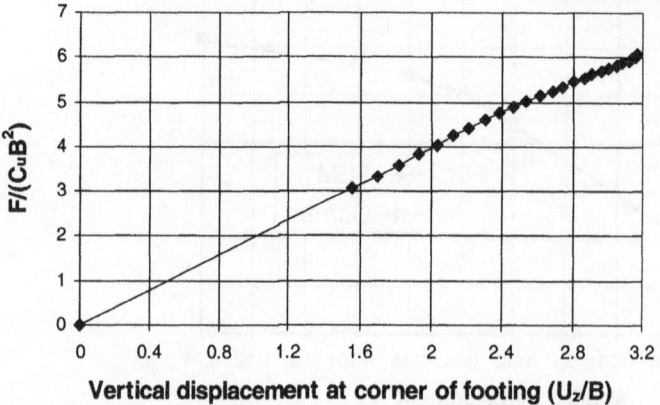

Figure 11.13: Load–(corner) settlement response for a square footing.

Figures 11.13 and 11.14 are load–displacement plots for the corner and the "mean" displacement. The latter has been approximated using the equation (Fox, 1948) $u_M = \frac{1}{3}(u_{corner} + 2u_{center})_{flexible}$. This should yield approximately the same displacement as a rigid footing. The displacements in these figures should be divided by 10^3 to recover the actual displacements. Although the exact solution to this problem is not known, the normalized collapse load for a rigid circular footing under the same condition is approximately 6, and it is probable that the collapse load for a square footing will not be much greater. With this discretization scheme, collapse is clearly indicated at a normalized load of just under 6.19 and the solution scheme is stable up to collapse.

The development of the yield zone is evident in Fig. 11.15, which identifies the yielded nodes (at a load level near to collapse) in plan and transverse sections. One might observe that the yield region extends laterally to encompass an area equal to at least four times the footing area and to a depth rather greater than the footing width.

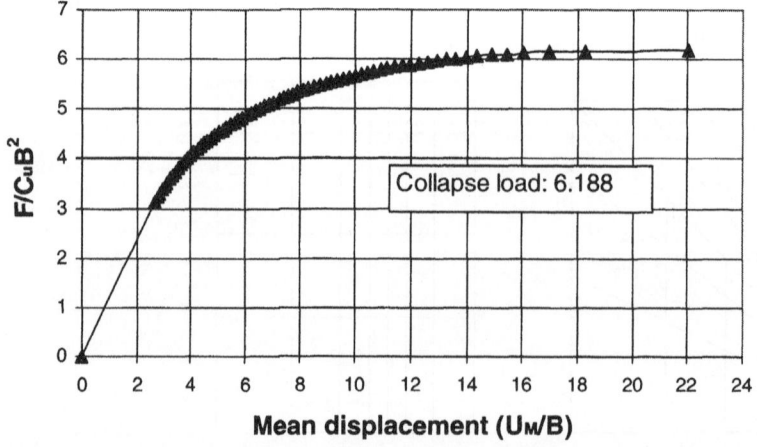

Figure 11.14: Load–(mean) settlement response for a square footing.

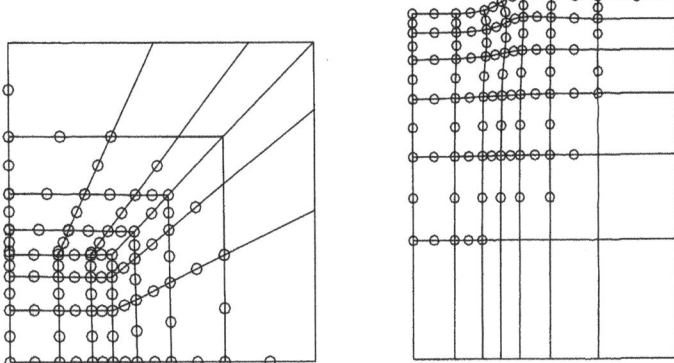

Figure 11.15: Deformed mesh and yielded nodes (just before collapse).

11.6 Multiplanar Tubular DX-Joint

In structural engineering, three-dimensional space frames containing details such as the multiplanar tubular DX-Joint depicted in Fig. 11.16 cannot be analyzed entirely satisfactorily by conventional methods. Subtle differences in detail, and the stress concentrations they engender, introduce uncertainties that only advanced numerical techniques can minimize. However, it would be fair to remark that the high surface to volume ratio of this particular (hollow tube) structural detail is not ideally suited for attack by general-purpose boundary element methods. Nevertheless, we present here a sample analysis, principally to demonstrate that the curved geometry of the problem presents no obstacle to the method.

The structural element (DX-Joint) consists of a large-diameter tube (chord) intersected orthogonally by two smaller diameter tubes (braces). The outer radii of the chord and braces are 228.6 mm and 28.58 mm, respectively, and the inner radii are half these values. For simplicity, only a short section of the tubes is analyzed and the eight-fold symmetry is exploited (Fig. 11.17). Even so, a large number of elements is needed: over 1,600 linear boundary elements and 1,000 cells were employed, defined by nearly 2,000 nodes, resulting in over 5,000 degrees of

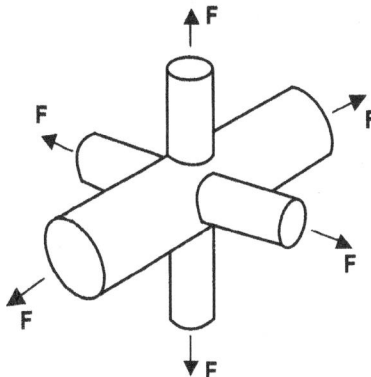

Figure 11.16: Multiplanar tubular DX-Joint subject to axial load.

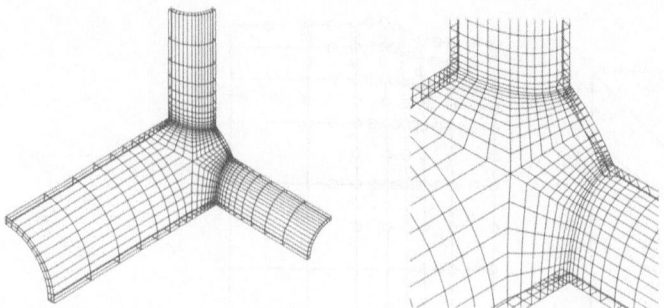

Figure 11.17: Boundary discretization of DX-Joint, with hidden detail removed.

freedom. Of the 2,000 nodes, only 300 were internal nodes, reflecting the high surface to volume ratio alluded to earlier.

A Von Mises (perfectly plastic) material is assumed, with the following parameters: Young's modulus $E = 200$ GPa, Poisson's ratio $\nu = 0.3$, and yield stress $\sigma_y = 0.4$ GPa. The uniformly distributed tensile stress F is applied to each end of the chord and braces. Full details of the input and output data are contained in the files DXJOINT.DAT and DXJOINT.OUT. Figure 11.18 is a plot of the displacement response of the end of the chord to increasing applied tensile stress. According to this analysis, collapse occurs with little warning, when the tensile stress is 0.234 GPa, although yielding begins much earlier. This is a somewhat higher collapse stress level than might be expected from general principles, and it should be confirmed by appropriate convergence studies.

The concentration of yielding at the brace–chord intersections is apparent in Fig. 11.19, which shows the locations of the yielded surface nodes at two stress levels (0.156 GPa and 0.234 GPa). Yielding begins in a narrow band at the root of the intersection, but collapse occurs only after the yield zone extends completely around the chord itself.

Equivalent stress contours, just before collapse, are plotted in Fig. 11.20. These highlight the stress concentration at the brace–chord intersections. From the stress data, the stress concentration factor (defined as the ratio of the absolute maximum principal stress to the applied stress) is 3.265. Also, the maximum equivalent plastic strain, which also occurs at the brace–chord intersection, is found to be 0.71%.

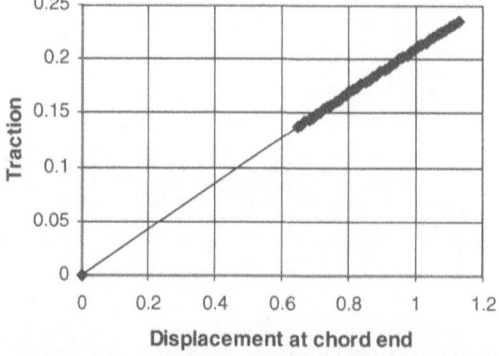

Figure 11.18: Stress–displacement response of chord.

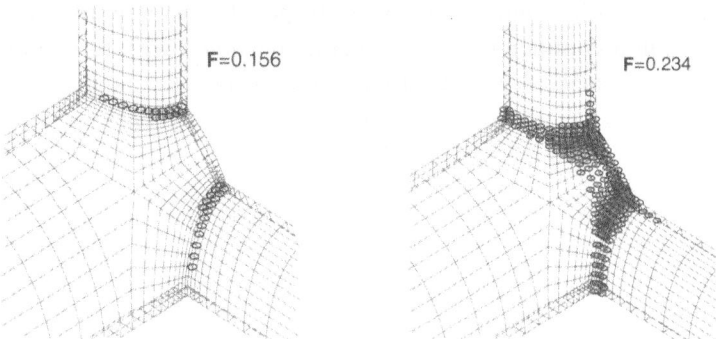

Figure 11.19: Evolution of yielding at brace–chord intersection.

11.7 Closure

In this chapter, we have sought to demonstrate that the boundary element method offers numerical modelers a practical means of solving realistic nonlinear problems in solid mechanics. All of the results obtained here have been obtained using very modest computing resources (i.e., Pentium™ II processors and single-precision

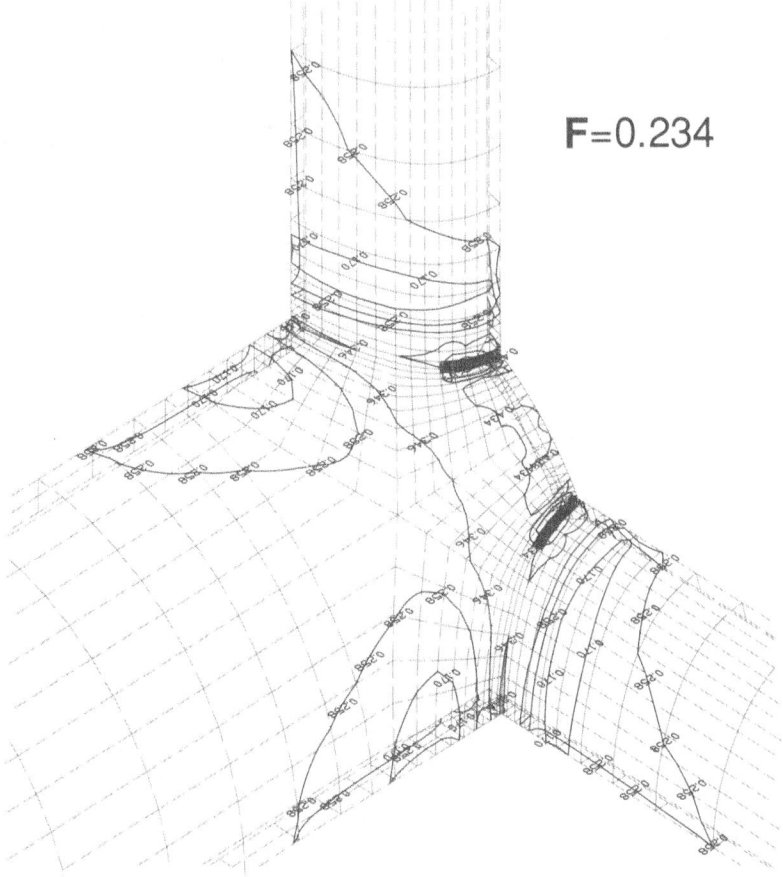

Figure 11.20: Contours of equivalent stress, just before collapse.

arithmetic), which suggests how much more might be possible on more powerful platforms. Naturally, in a commercial environment, one would wish to employ more advanced features, such as enhanced input–output facilities, multiregion capabilities, and a more efficient matrix solver for large-scale problems. Some further developments of program BEMECH that address these issues, and others, are discussed in the following chapter.

CHAPTER TWELVE

Epilogue

12.1 Review

Computer programs must evolve or become extinct as fitter competitors take their place. Indeed, it might well be argued that the battle is being fought on a wider scale between finite element methods and boundary element methods. Thus, part of our motivation in publishing this work has been to provide a fillip to those who would wish to redress the competitive advantage that the former enjoys, by virtue of its long-established position in the world of numerical modeling. We have tried to demonstrate that boundary element methods are now sufficiently powerful to match finite element methods in certain applications, particularly in three-dimensional analyses, and indeed to surpass them in terms of data processing efficiency and numerical accuracy. Of course, boundary element methods have far wider applications in mechanics, and other areas in the physical sciences, than we have been able to discuss here, and these are well documented in a number of texts (e.g., Banerjee, 1994). What emerges from a study of these applications is that, in many cases, there is a remarkable unity in the formulation of the boundary integral equations. Hence, the formulations and numerical algorithms described in this book have far wider applications than may be apparent at first sight. In particular, the numerical integrations of kernel functions are universal in boundary element methods. But, returning to our opening theme, we emphasize the program BEMECH is not intended to be, and could not be, the final result of some long evolutionary process. The version published here is in fact an offspring of the main line, shorn of some attributes of its parent and stylized for a general audience. Several code enhancements, some of which are currently under development, are discussed in the remainder of this chapter.

12.2 The Way Forward

Several suggestions for further developments of the BEMECH code intended to improve computational efficiency and enhance the program's functions are described below.

12.2.1 Automatic Integration

The adaptive (automatic) integration algorithm described in Section 4.5 uses element subdivision whenever Gauss integration over a complete element becomes uneconomic. The Gauss integration (order) over each subelement is determined automatically and varies from subelement to subelement. However, the size of each subelement is the same and greater computational efficiency could be realized if a more sophisticated algorithm was employed. Such an algorithm would generate subelements of optimal size (graded from the smallest one nearest the singularity to the largest one furthest away).

12.2.2 Computation of Boundary Stresses

The traction recovery method used in BEMECH can yield inaccurate boundary stresses because it involves displacement gradients. The direct evaluation of boundary stresses (e.g., Guiggiani et al., 1992; Huber et al., 1996; Poon, Mukherjee, & Bonnet 1998) offers the prospect of more accurate results, albeit at the cost of solving a very large set of equations. Provided that computer memory and computational time are not inhibiting constraints, the direct method should become the method of choice.

12.2.3 Stress-Return Algorithm

When large load increments are imposed, computed stress paths may deviate significantly from the true paths. To counter this problem, in the context of finite element methods, Ortiz & Simo (1986) and Simo & Govindjee (1991) proposed a stress-return algorithm, which Gao (1999) has incorporated within a boundary element algorithm. In this technique, the excess stress (at each node) is drawn back to the yield surface using a local iteration procedure. We assume that for a given strain ε, after the kth iteration in the current increment, the total stress–strain Eq. (7.20) is not exactly satisfied. The residual R^k is then

$$
\begin{aligned}
R^k &= \sigma^k - \hat{\sigma}(\varepsilon, \varepsilon^{\mathrm{p}}, h^\alpha)^k \\
&= \sigma^k - \hat{\sigma}(\varepsilon, \lambda^k)
\end{aligned}
\tag{12.1}
$$

To reduce the residual to zero, the values of σ and λ must be corrected in the $(k+1)$-th iteration, that is,

$$
\begin{aligned}
\sigma^{k+1} &= \sigma^k + \dot{\sigma} \\
\lambda^{k+1} &= \lambda^k + \dot{\lambda}
\end{aligned}
\tag{12.2}
$$

Now, the truncated Taylor's series expansion of Eq. (12.1), noting that ε is fixed, yields

$$
\begin{aligned}
R^{k+1} &= R^k + \frac{\partial R}{\partial \sigma^k} : \dot{\sigma} + \frac{\partial R}{\partial \lambda^k}\dot{\lambda} \\
&= R^k + \dot{\sigma} - \frac{\partial \hat{\sigma}}{\partial \lambda^k}\dot{\lambda}
\end{aligned}
\tag{12.3}
$$

From the flow rule (Eq. 7.28), we have

$$\frac{\partial \hat{\sigma}}{\partial \lambda^k} = -\mathbf{D}^e : \frac{\partial f}{\partial \sigma^k} \tag{12.4}$$

Substituting this equation into Eq. (12.3) and forcing $R^{k+1} = 0$ leads to

$$\dot{\sigma} = -R^k - \mathbf{D}^e : \frac{\partial f}{\partial \sigma^k} \lambda \tag{12.5}$$

Similarly, the residual f^k of the yield function (Eq. 7.19) can be written for the kth iteration as

$$f^k = f(\sigma, \varepsilon^p, h^\alpha)^k \tag{12.6}$$

The Taylor's series expansion leads to

$$f^{k+1} = f^k + \frac{\partial f}{\partial \sigma^k} : \dot{\sigma} + \frac{\partial f}{\partial \varepsilon^p} : \dot{\varepsilon}^p + \frac{\partial f}{\partial h^\alpha} h^\alpha \tag{12.7}$$

Substituting Eqs. (7.32) and (7.36) into this equation, we get:

$$f^{k+1} = f^k + \frac{\partial f}{\partial \sigma^k} : \dot{\sigma} - \Gamma \lambda \tag{12.8}$$

where Γ is determined using Eq. (7.46). Now, forcing $f^{k+1} = 0$, solving for λ, and making use of Eq. (12.5) yields

$$\lambda - \frac{1}{\Psi} \left(f^k - \frac{\partial f}{\partial \sigma^k} : R^k \right) \tag{12.9}$$

where Ψ is determined using Eq. (7.41). The stress changes can then be found by substituting this value of λ into Eq. (12.5). The stresses and the plastic multiplier are then updated using Eq. (12.2). This procedure is then repeated until both the residuals are sufficiently small. In BEMECH, it would be encoded in subroutine P_M_ITER at the end of each load increment.

12.2.4 System Equation Solver

In BEMECH, the Gauss–Jordan elimination method is used to solve the system equations. But for large problems, this method is inefficient and may yield inaccurate results owing to the accumulation of round-off error. Iterative solution techniques (such as the GMRES algorithm; Saad & Schultz, 1986) have begun to gain favor (e.g., Leung & Walker, 1997). The GMRES algorithm operates directly on the original set of unsymmetric matrix equations and requires only one matrix–vector multiplication in each iteration.

12.2.5 Local Boundary Conditions

In some cases (Fig. 12.1), boundary conditions are specified, at least in part, in terms of some local system of axes. In this example, the "roller" boundary conditions at lower left must be specified in terms of a local Cartesian coordinate system

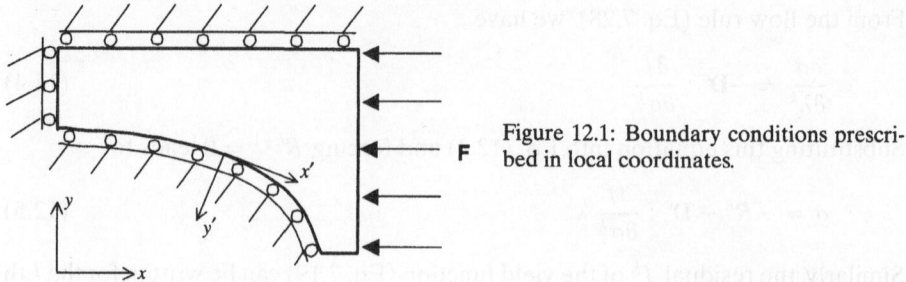

Figure 12.1: Boundary conditions prescribed in local coordinates.

(x', y') directed along the normal and tangential directions to the curved boundary. Consequently, the final system equations must be cast in terms of these local displacements and tractions, and this implies that the kernel functions themselves (in subroutine EVAL_HG) must be expressed in terms of these local coordinates.

12.2.6 Nonlinear Hardening

In BEMECH, the hardening parameters are assumed to be constants, but in general most materials exhibit behavior that can be better modeled if the hardening parameters are assumed to be functions of equivalent plastic strain. A simple way to accommodate materials of this type is to assume that the hardening is piecewise linear. A more rigorous, albeit more complicated, method is to interpolate using spline functions (Gao & Zhong, 1992). These changes would affect subroutines P_M_ITER, SIG_SCALE, and DF_MATRX.

12.2.7 Advanced Yield Functions

More sophisticated yield criteria than those incorporated within BEMECH may be desirable in many cases. Yield criteria such as those described by Lade (1977) and Ottosen (1977), for example, can be very easily implemented, once the coefficients C_1, C_2, and C_3 (refer to Section 7.5.3) are determined, by simply modifying subroutine DF_DSIG. More complex models, such as the "critical state" models for soils, could be included too, without much difficulty.

12.2.8 Finite Strain Elasto–Plasticity

The constitutive relationships developed in Chapter Seven are applicable in finite plasticity provided that the stress tensor is understood to be the symmetric Piola–Kirchhoff stress tensor. It turns out that if an updated Lagrangian description is employed, and elastically isotropic and elastically homogeneous material is assumed, the boundary integral equations for finite plasticity (Okada & Atluri, 1994) become very similar to the equations for infinitesimal plasticity. The difference is that the domain integrals involve the Cauchy stresses and the gradients of the displacements, in addition to the initial stress rates. The Cauchy stresses can be determined during incremental iteration and the displacement gradients can be evaluated using either of two techniques. The first of these employs interior displacement gradient integral equations, which are similar to the interior stress integral equations. Like the stresses, the displacement gradients are hypersingular at the boundary, so a technique analogous to the traction recovery method must be employed to obtain them. Alternatively, the displacement gradients may be determined using the "full tangent stiffness field-boundary element method." In this technique, both the boundary and interior displacement gradients are evaluated by differentiating the nodal shape functions and the resultant system equations involve the displacements inside and on the boundary of the body as the primary variables. This technique is easily coded.

12.2.9 Infinite Boundary Elements

Half-space problems, where an infinitely extended surface is traction free, except perhaps in the near field, are quite commonly encountered in practice. Enforcing this condition, without resorting to either a half-space fundamental solution or discretizing the surface, is best done using "infinite elements," a concept borrowed from the finite element literature. In the boundary element context, infinite elements were first developed by Watson (1979). More recently, significant efficiency gains have been realized by Beer & Watson (1989), Davies & Bu (1996), and Gao & Davies (1998). In the last of these, the infinite surface integrals are transformed into line integrals that can be evaluated analytically, which yields a particularly simple and efficient algorithm.

12.2.10 Multiple Regions

Multiple-region capability is obviously needed to analyze problems that involve materials consisting of several contiguous zones of different properties. Multiple-region algorithms are also useful in other instances too, for example, where artificial subdivision of a slender region may improve the conditioning of the system equations. For a large complex region, such subdivision may be justified on the grounds that the resulting block-banded system equations can be generated and solved more quickly than the fully populated original set. Some workers (Blandford, Ingraffea, & Liggett 1981) have also attacked fracture as a multiple-region problem, with an interface centered on the crack. One method of assembly of the global system of equations is discussed briefly in Chapter Four; other methods are reviewed by Gao & Davies (2000a). An important consideration is that rounding off of corners and edges is impossible in multiple-region problems and, unless discontinuous elements are employed, the methods described in Chapter Four become imperative.

APPENDIX A

Derivation of Kernel Functions

A.1 Derivation of the Strain Kernel

For simplicity, the Kelvin fundamental solution (Chapter 3) can be rewritten in the form

$$u_j = U_{ij} e_i \tag{A.1}$$

where u_j is the jth Cartesian component of displacement and e_i is the unit point force in the ith Cartesian direction. Now, recalling the strain–displacement relationships from Chapter Two,

$$\varepsilon_{jk} = (u_{j,k} + u_{k,j})/2 \tag{A.2}$$

we obtain

$$\varepsilon_{jk} = E_{ijk} e_i \tag{A.3}$$

where the strain kernel E_{ijk} is defined by the equation

$$E_{ijk} = (U_{ij,k} + U_{ik,j})/2 \tag{A.4}$$

The spatial differentiation is carried out with respect to the field point q because this is the point at which displacements are calculated. To differentiate the fundamental solution, the following results are required:

$$\left(\frac{1}{r}\right)_{,j} = \frac{\partial}{\partial x_j}\left(\frac{1}{r}\right) = \frac{\partial}{\partial r}\left(\frac{1}{r}\right)\frac{\partial r}{\partial x_j}$$
$$= \frac{-1}{r^2}\frac{r_j}{r} = \frac{-r_j}{r^3} \tag{A.5}$$

Similarly,

$$\left(\frac{1}{r^3}\right)_{,j} = \frac{-3r_j}{r^5} \tag{A.6}$$

Also, note that

$$r_{i,j} = \delta_{ij} \tag{A.7}$$

and

$$r,_i = r_i/r \tag{A.8}$$

Now, from Eq. (A.1) or, more precisely, its explicit form in Chapter Three, we have

$$U_{ij} = A\left[B\frac{\delta_{ij}}{r} + \frac{r_i r_j}{r^3} \right] \tag{A.9}$$

where

$$\begin{aligned} A &= 1/[16\pi\,G(1 - \nu)] \\ B &= 3 - 4\nu \end{aligned} \tag{A.10}$$

Making use of Eqs. (A.5)–(A.10), we obtain

$$U_{ik,j} = \frac{A}{r^3}[-B\delta_{ik}r_j + r_i\delta_{kj} + r_k\delta_{ij} - 3r_i r_j r_k/r^2] \tag{A.11}$$

Switching subscripts, we immediately obtain

$$U_{ij,k} = \frac{A}{r^3}[-B\delta_{ij}r_k + r_i\delta_{jk} + r_j\delta_{ik} - 3r_i r_j r_k/r^2] \tag{A.12}$$

Hence, using Eq. (A.4), we obtain the result quoted in Chapter Three:

$$E_{ijk} = \frac{-A}{r^2}[C(\delta_{ik}r,_j + \delta_{ij}r,_k) - \delta_{jk}r,_i + 3r,_i\,r,_j\,r,_k] \tag{A.13}$$

where

$$C = 1 - 2\nu \tag{A.14}$$

A.2 Derivation of the Stress Kernel

The stress kernel Σ_{ijk} is defined in Chapter Three by the equation

$$\sigma_{jk} = \Sigma_{ijk}e_i \tag{A.15}$$

where e_i is the ith Cartesian component of the point force system. Hooke's law, relating stresses and strains, is defined in Chapter Two by the equation

$$\sigma_{jk} = \lambda\delta_{jk}\varepsilon_{mm} + 2G\varepsilon_{jk} \tag{A.16}$$

Because Lame's constant λ is linked to the shear modulus G by the equation

$$\lambda = 2G\nu/(1 - 2\nu) \tag{A.17}$$

we can rewrite Hooke's law in the form

$$\sigma_{jk} = 2G\left[\frac{\nu}{1 - 2\nu}\delta_{jk}\varepsilon_{mm} + \varepsilon_{jk} \right] \tag{A.18}$$

Therefore, using Eqs. (A.3), (A.15), and (A.18), we obtain

$$\Sigma_{ijk} = 2G\left[\frac{v}{1-2v}\delta_{jk}E_{imm} + E_{ijk}\right] \tag{A.19}$$

Noting that $\delta_{mm} = 3$, $r_{,m}r_{,m} = 1$, and $\delta_{mi}r_{,m} = r_{,i}$ we now set $j = k = m$ in Eq. (A.13) to obtain

$$E_{imm} = -2ACr_{,i}/r^2 \tag{A.20}$$

Finally, substituting this equation and Eq. (A.13) into Eq. (A.19), we obtain the result quoted in Chapter Three, namely,

$$\Sigma_{ijk} = \frac{-2GA}{r^2}[C(\delta_{ik}r_{,j} + \delta_{ij}r_{,k} - \delta_{jk}r_{,i}) + 3r_{,i}r_{,j}r_{,k}] \tag{A.21}$$

A.3 Derivation of the Traction Kernel

The traction kernel is defined in Chapter Three by the equation

$$t_j = T_{ij}e_i \tag{A.22}$$

where t_j are traction components (on a plane with outward normal direction cosines n_k) resulting from point forces e_i. Because the tractions are related to the stress state by the equilibrium condition (Chapter Two)

$$t_j = \sigma_{jk}n_k \tag{A.23}$$

then, using Eq. (A.15), we obtain

$$T_{ij} = \Sigma_{ijk}n_k \tag{A.24}$$

Noting such contractions as $\delta_{jk}r_{,i}n_j = r_{,i}n_k$, etc., we obtain the result quoted in Chapter Three, namely,

$$T_{ij} = \frac{-2GA}{r^2}[C(n_ir_{,j} - n_jr_{,i}) + (3r_{,i}r_{,j} + C\delta_{ij})n_mr_{,m}] \tag{A.25}$$

The repeated (dummy) subscript in the last term is arbitrary, as emphasized by the use of the subscript m here.

A.4 Kernel Functions for Plane Strain and Plane Stress

By integrating Kelvin's solution for a line source $e_k(p)$, where $k = 1$ and 2 only, along the x_3-axis, we obtain the displacement kernel for plane strain, namely,

$$U_{ij} = -2A[B\delta_{ij}\log_e(r) - r_{,i}r_{,j}] \tag{A.26}$$

where, as before, the constants are

$$\begin{aligned} A &= 1/[16\pi G(1-v)] \\ B &= 3 - 4v \end{aligned} \tag{A.27}$$

Naturally, in this equation, the subscripts i and j range from 1 to 2 only. Following the same procedure demonstrated above for the three-dimensional case, here substituting Eq. (A.26) into Eq. (A.4), we obtain the strain kernel:

$$E_{ijk} = \frac{-2A}{r}[C(\delta_{ik}r_{,j} + \delta_{ij}r_{,k}) - \delta_{jk}r_{,i} + 2r_{,i}\,r_{,j}\,r_{,k}] \qquad (A.28)$$

where

$$C = 1 - 2v \qquad (A.29)$$

Likewise, the stress kernel is found to be

$$\Sigma_{ijk} = \frac{-4GA}{r}[C(\delta_{ik}r_{,j} + \delta_{ij}r_{,k} - \delta_{jk}r_{,i}) + 2r_{,i}\,r_{,j}\,r_{,k}] \qquad (A.30)$$

and the traction kernel is

$$T_{ij} = \frac{-4GA}{r}[C(n_{i}r_{,j} - n_{j}r_{,i}) + (2r_{,i}\,r_{,j} + C\delta_{ij})n_{m}r_{,m}] \qquad (A.31)$$

To determine the internal stresses under plane strain conditions, resulting from known boundary tractions and displacements, the kernel functions U_{ijk} and T_{ijk} (defined in Chapter Three) are required. Following the procedure outlined there, we get

$$U_{ijk} = \frac{1}{4\pi(1-v)}\frac{1}{r}[C(\delta_{ki}r_{,j} + \delta_{kj}r_{,i} - \delta_{ij}r_{,k}) + 2r_{,i}\,r_{,j}\,r_{,k}] \qquad (A.32)$$

and

$$T_{ijk} = \frac{G}{2\pi(1-v)}\frac{1}{r^{2}}\{2r_{,m}\,n_{m}[C\delta_{ij}r_{,k} + v(\delta_{ik}r_{,j} + \delta_{jk}r_{,i}) - 4r_{,i}\,r_{,j}\,r_{,k}]$$

$$+ 2v(n_{i}r_{,j}\,r_{,k} + n_{j}r_{,i}\,r_{,k}) + C(2n_{k}r_{,i}\,r_{,j} + n_{j}\delta_{ik} + n_{i}\delta_{jk}) - Dn_{k}\delta_{ij}\} \qquad (A.33)$$

where

$$D = 1 - 4v \qquad (A.34)$$

For plane stress, the corresponding results can be obtained simply by replacing the Poisson's ratio by the "equivalent" value, $v/(1+v)$. Moreover, if the kernel functions for three dimensions and for two dimensions are compared, it will be observed that they are very similar. For programming purposes, this similarity can be exploited to streamline the code.

Shape Functions

The Serendipity shape functions (Zienkiewicz & Taylor, 1989) for linear and quadratic interpolation in one, two, and three dimensions are listed below. The nodes and their intrinsic coordinates $(\xi_\alpha, \eta_\alpha)$ are defined in the figures. Here, we adopt the usual convention that the intrinsic coordinates take values in the range ± 1, although alternatives (such as the range 0–1) are equally possible and may be advantageous in some circumstances. We also restrict ourselves to the prismatic family of elements, although shape functions for triangular and tetrahedral elements are readily available in the literature. Perhaps it is worth restating that in a boundary element formulation in n-dimensional space, only shape functions for the $(n-1)$ dimension are needed. Thus, the n-dimensional shape functions are only required if one of the domain integrals cannot be conveniently transformed into a boundary integral, as for example in elasto–plastic flow problems.

B.1 One-Dimensional Shape Functions

For linear interpolation, two nodes must be defined as depicted in Fig. B.1. The shape functions for these two nodes can be expressed concisely as

$$N_\alpha(\xi) = \frac{1}{2}(1 + \xi_\alpha \xi) \tag{B.1}$$

Alternatively, in explicit form, they are

$$N_1(\xi) = \frac{1}{2}(1 - \xi)$$

$$N_2(\xi) = \frac{1}{2}(1 + \xi) \tag{B.2}$$

For quadratic interpolation, an additional node $(\alpha = 3)$ must be specified at the

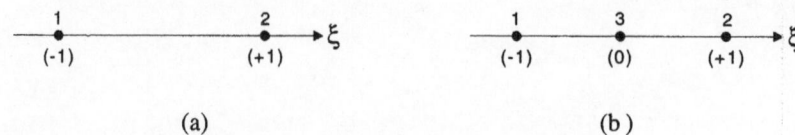

(a)   (b)

Figure B.1: Serendipity line elements: (a) linear, (b) quadratic.

center of the intrinsic element. The shape functions are

$$N_1(\xi) = \frac{1}{2}\xi(\xi - 1)$$

$$N_2(\xi) = \frac{1}{2}\xi(1 + \xi) \tag{B.3}$$

$$N_3(\xi) = 1 - \xi^2$$

B.2 Two-Dimensional Shape Functions

For linear interpolation, four nodes must be defined as depicted in Fig. B.2. Denoting the node numbers by the subscript α, we can express all four shape functions ($\alpha = 1, 2, 3, 4$) in the form

$$N_\alpha(\xi, \eta) = \frac{1}{4}(1 + \xi_\alpha\xi)(1 + \eta_\alpha\eta) \tag{B.4}$$

where $(\xi_\alpha, \eta_\alpha)$ denotes the intrinsic coordinates of the αth node. Alternatively, in explicit form, the shape functions are

$$N_1(\xi,\eta) = \frac{1}{4}(1 - \xi)(1 - \eta)$$

$$N_2(\xi,\eta) = \frac{1}{4}(1 + \xi)(1 - \eta)$$

$$N_3(\xi,\eta) = \frac{1}{4}(1 + \xi)(1 + \eta) \tag{B.5}$$

$$N_4(\xi,\eta) = \frac{1}{4}(1 - \xi)(1 + \eta)$$

(a)   (b)

Figure B.2: Serendipity quadrilateral elements: (a) linear, (b) quadratic.

For quadratic interpolation, an additional four nodes must be specified – at the center of each side of the intrinsic element. The shape functions for the corner nodes ($\alpha = 1, 2, 3, 4$) now become

$$N_\alpha(\xi,\eta) = \frac{1}{4}(1 + \xi_\alpha\xi)(1 + \eta_\alpha\eta)(-1 + \xi_\alpha\xi + \eta_\alpha\eta) \tag{B.6}$$

and those for the mid-side nodes ($\alpha = 5, 6, 7, 8$) are

$$N_\alpha(\xi,\eta) = \frac{1}{2}(1 + \xi_\alpha\xi + \eta_\alpha\eta)\{1 - (\xi_\alpha\eta)^2 - (\eta_\alpha\xi)^2\} \tag{B.7}$$

As explained in the text, these shape functions cannot describe the complete quadratic, which consists of nine terms.

B.3 Three-Dimensional Shape Functions

For linear interpolation, eight nodes must be defined as depicted in Fig. B.3. Denoting the node numbers by the subscript α, we can express all eight shape functions ($\alpha = 1$–8) in the form

$$N_\alpha(\xi_1, \xi_2, \xi_3) = \frac{1}{8}\left(1 + \xi_1^\alpha\xi_1\right)\left(1 + \xi_2^\alpha\xi_2\right)\left(1 + \xi_3^\alpha\xi_3\right) \tag{B.8}$$

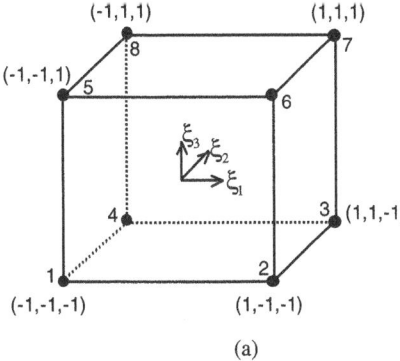

(a)

Figure B.3: Serendipity brick elements: (a) linear, (b) quadratic.

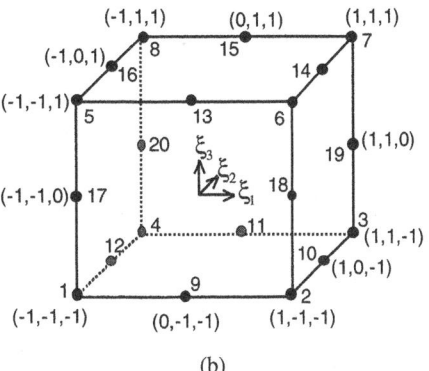

(b)

where $(\xi_1^\alpha, \xi_2^\alpha, \xi_3^\alpha)$ denotes the intrinsic coordinates of the αth node. For quadratic interpolation in three dimensions, the "brick" element is defined by twenty nodes. For the eight corner nodes ($\alpha = 1$–8), the shape functions are

$$N_\alpha(\xi_1, \xi_2, \xi_3) = \frac{1}{8}\left(1 + \xi_1^\alpha \xi_1\right)\left(1 + \xi_2^\alpha \xi_2\right)\left(1 + \xi_3^\alpha \xi_3\right)\left(\xi_1^\alpha \xi_1 + \xi_2^\alpha \xi_2 + \xi_3^\alpha \xi_3 - 2\right)$$

$$\text{(B.9)}$$

and for the twelve remaining mid-side nodes ($\alpha = 9$–20), the shape functions are

$$N_\alpha(\xi_1, \xi_2, \xi_3) = \frac{1}{4}\left(1 + \xi_1^\alpha \xi_1\right)\left(1 + \xi_2^\alpha \xi_2\right)\left(1 + \xi_3^\alpha \xi_3\right)$$

$$\times \left\{1 + \left[(\xi_1^\alpha)^2 - 1\right]\xi_1^2 + \left[(\xi_2^\alpha)^2 - 1\right]\xi_2^2 + \left[(\xi_3^\alpha)^2 - 1\right]\xi_3^2\right\}$$

$$\text{(B.10)}$$

APPENDIX C

Degenerate Elements: Singular Mapping

Certain transformations from global to local coordinate systems are routinely employed in the boundary element method to nullify weak singularities. The Jacobian of the transformation J_c between these coordinate systems describes the degree to which the space is warped by these transformations and a short trek through some of the mathematics will be useful. Here, we consider only the two-dimensional case in detail. In Gauss quadrature, we integrate in an intrinsic coordinate system, which requires a transformation from the global (x, y) system to the local (ξ_1, ξ_2) system. This transformation yields the equivalence

$$d\Omega(x, y) = J_c d\Omega(\xi_1, \xi_2) \tag{C.1}$$

where the Jacobian J_c is

$$
\begin{aligned}
J_c &= \begin{vmatrix} \dfrac{\partial x}{\partial \xi_1} & \dfrac{\partial y}{\partial \xi_1} \\[2mm] \dfrac{\partial x}{\partial \xi_2} & \dfrac{\partial y}{\partial \xi_2} \end{vmatrix} \\[4mm]
&= \frac{\partial x}{\partial \xi_1}\frac{\partial y}{\partial \xi_2} - \frac{\partial x}{\partial \xi_2}\frac{\partial y}{\partial \xi_1}
\end{aligned}
\tag{C.2}
$$

This transformation is commonly employed to map a quadrilateral cell in the global system into a square cell in the local system (Fig. C.1).

Referring to Fig. C.1, through any point $q(\xi_1, \xi_2)$, we can define two lines L_1^ξ and L_2^ξ directed along the intrinsic coordinate axes (i.e., along $\xi_2 = constant$ and $\xi_1 = constant$, respectively). These are curves in global space but orthogonal lines in the intrinsic system. Now, we define a local non-Cartesian coordinate system (Fig. C.2) with its origin at $q(x, y)$ and axes (x', y') directed along L_1 and L_2, respectively. The coordinate transformation from the (x', y') system to the (x, y) system is

$$
\begin{aligned}
x &= x' \cos\theta_1 + y' \sin\theta_2 \\
y &= x' \sin\theta_1 + y' \cos\theta_2
\end{aligned}
\tag{C.3}
$$

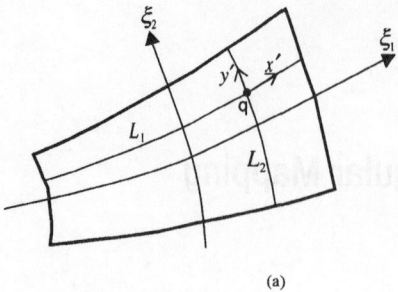

(a)

Figure C.1: Mapping of a two-dimensional cell.

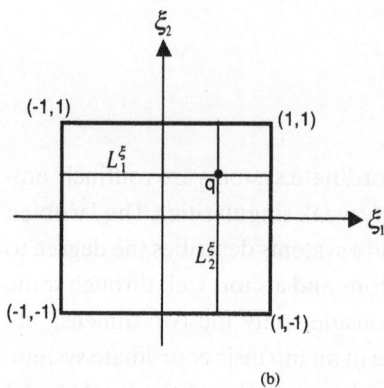

(b)

The Jacobian of the transformation between these two systems is

$$d\Omega(x, y) = dx \, dy$$
$$= J_L \, dx' dy' \qquad\qquad (C.4)$$

where J_L is, from Eq. (C.2),

$$J_L = \begin{vmatrix} \cos\theta_1 & \sin\theta_2 \\ \sin\theta_1 & \cos\theta_2 \end{vmatrix} \qquad\qquad (C.5)$$
$$= \cos\theta_1 \cos\theta_2 - \sin\theta_1 \sin\theta_2$$

Now, let us assume that the lines L_1^ξ and L_2^ξ are divided into m segments, and

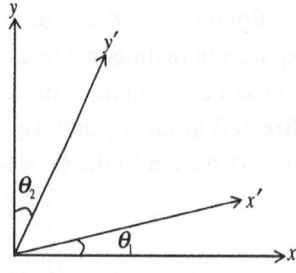

Figure C.2: Global (Cartesian) and local (non-Cartesian) coordinate systems.

likewise their counterparts in global space. In the limit, we have

$$\Delta\xi_1 = L_1^\xi/m = d\xi_1$$

$$\Delta\xi_2 = L_2^\xi/m = d\xi_2$$

$$\Delta x' = \Delta L_1 = L_1/m = dx'$$ (C.6)

$$\Delta y' = \Delta L_2 = L_2/m = dy'$$

Substituting Eq. (C.6) into Eq. (C.4), we obtain

$$d\Omega(x, y) = \frac{J_L L_1 L_2}{m^2}$$ (C.7)

For the intrinsic coordinate system, clearly

$$d\Omega(\xi_1, \xi_2) = d\xi_1 d\xi_2 = \frac{L_1^\xi L_2^\xi}{m^2}$$ (C.8)

Because, in the intrinsic coordinate system, the cell's dimensions are two units of length along the ξ_1 and ξ_2 directions, that is,

$$L_1^\xi \equiv 2$$

$$L_2^\xi \equiv 2$$ (C.9)

then, substituting Eqs. (C.7) and (C.8) into Eq. (C.1), we obtain

$$J_c = \frac{J_L L_1 L_2}{4}$$ (C.10)

From this equation, we can see that the Jacobian is proportional to the lengths of the lines L_1 and L_2 through the point q. For a degenerate cell (Fig. C.3), when the point q approaches the vertex of the triangle and L_2 approaches zero, then clearly the Jacobian approaches zero too.

To determine the nature of the relationship between the distance r of the field point q to the degenerate vertex (at p) and the Jacobian J_c, we denote the angle (at q) between the vector $\vec{qp}$ and L_2 by the symbol α (refer to Fig. C.3). For a small increment ΔL_2, we have

$$r\Delta\theta = \Delta L_2 \sin\alpha$$ (C.11)

Dividing the angle θ (in Fig. C.3) into m parts, we define

$$\Delta\theta = \theta/m$$ (C.12)

Figure C.3: Two-dimensional degenerate cell.

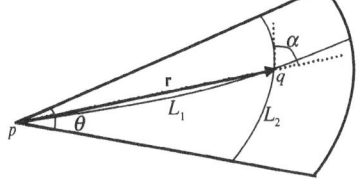

Substituting Eqs. (C.11), (C.12), and (C.6) into Eq. (C.4), we obtain

$$
\begin{aligned}
d\Omega(x, y) &= J_L dx' dy' \\
&= \frac{J_L L_1 r\theta}{m^2 \sin \alpha}
\end{aligned}
\tag{C.13}
$$

Substituting this equation and Eq. (C.8) into Eq. (C.1), and making use of Eq. (C.9), we obtain

$$
J_c = \frac{J_L L_1 r\theta}{4 \sin \alpha}
\tag{C.14}
$$

From this equation, we observe that the Jacobian is proportional to r, and hence, within the boundary integral formulation, this mapping can be used to nullify the weak (r^{-1}) singularities in the two-dimensional kernel functions. Similarly, it may be shown that for a degenerate three-dimensional cell the Jacobian is proportional to r^2, and this is sufficient to nullify the corresponding weak singularity in three dimensions.

Elasto–Plastic Flow Theory

D.1 Derivation of the Plastic Flow Rule and Plastic Loading Rule

The Il'iushin postulate (Chapter 7) conjectures that the work done in a closed strain cycle is nonnegative, that is,

$$\oint \sigma : d\varepsilon = \int_{t_1}^{t_2} \sigma : \dot{\varepsilon} \, dt \geq 0 \tag{D.1}$$

Let ε^0 be some arbitrary strain inside the loading surface and consider a strain cycle beginning and ending at ε^0. For this cycle, noting that $\varepsilon(t_1) = \varepsilon(t_2) = \varepsilon^0$, we can write the inequality (D.1) (through integration by parts) as

$$\int_{t_1}^{t_2} (\varepsilon - \varepsilon^0) : \dot{\sigma} \, dt \leq 0 \tag{D.2}$$

Now, let us consider a typical stress–strain curve for uniaxial tests, such as that depicted in Fig. D.1. We consider a strain cycle in which the deformation takes place elastically from t_1 to a yield point t_y, then continues elasto–plastically to t_z, and finally unloads elastically to point t_2, that is,

$$t_1 \rightarrow t_y \rightarrow t_z \rightarrow t_2 \tag{D.3}$$

During this strain cycle, plastic deformation occurs only in the stage $t_y \rightarrow t_z$ and elastic unloading does not (of course) produce irreversible deformation. Incidentally, for nonlinear elastic materials, the elastic deformation paths, $t_1 \rightarrow t_y$ and $t_z \rightarrow t_2$, are not straight lines. Now, since t_y and t_z are the turning points between elastic and plastic deformations, we denote t_{y-} and t_{y+} as the left limit and right limit of the point t_y, respectively, and similarly we denote t_{z-} and t_{z+} as the left and right limits of the point t_z. The points t_{y+} and t_{z-} are on the loading surface. Differentiating Eq. (7.20) with respect to time leads to

$$\dot{\sigma} = (D - D^e) : \dot{\varepsilon} + D^e : \dot{\varepsilon} + \dot{G} \tag{D.4}$$

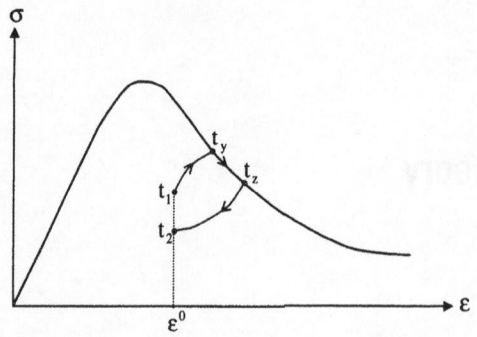

where

$$\dot{G} = \frac{\partial \hat{\sigma}}{\partial \varepsilon^p} : \dot{\varepsilon}^p + \frac{\partial \hat{\sigma}}{\partial h^\alpha} h^\alpha \tag{D.5}$$

and

$$D^e = \frac{\partial \hat{\sigma}}{\partial \varepsilon}(t_{y-}) \tag{D.6}$$

Since t_{y-} is located within the elastic region, the fourth-order tensor D^e is the "elastic constitutive tensor." Substituting Eq. (D.4) into Eq. (D.2), we obtain

$$\int_{t_1}^{t_2} (\varepsilon - \varepsilon^0) : (D - D^e) : \dot{\varepsilon} dt + \int_{t_1}^{t_2} (\varepsilon - \varepsilon^0) : D^e : \dot{\varepsilon} dt$$

$$+ \int_{t_1}^{t_2} (\varepsilon - \varepsilon^0) : \dot{G} dt \leq 0 \tag{D.7}$$

The second integral in this equation can be readily shown to be zero, because $\varepsilon(t_1) = \varepsilon(t_2) = \varepsilon^0$. The last integral can be written in the following form, since only in the interval $t_{y+} \to t_{z-}$ does the plastic strain ε^p and the internal variables h^α alter:

$$\int_{t_1}^{t_2} (\varepsilon - \varepsilon^0) : \dot{G} dt = \int_{t_{y+}}^{t_{z-}} (\varepsilon - \varepsilon^0) : \dot{G} dt \tag{D.8}$$

Expanding this equation, using Taylor's series about the point t_{y+}, yields

$$\int_{t_1}^{t_2} (\varepsilon - \varepsilon^0) : \dot{G} dt = (\varepsilon - \varepsilon^0) : \dot{G} \Delta t + [\dot{\varepsilon} : \dot{G} + (\varepsilon - \varepsilon^0) : \ddot{G}] \frac{(\Delta t)^2}{2} + O(\Delta t)^3 \tag{D.9}$$

in which $\Delta t = t_{z-} - t_{y+} = t_z - t_y$ and $O(\Delta t)^3$ represents the infinitesimal terms of third and higher orders of Δt as $\Delta t \to 0$. Similarly, the Taylor's series expansion of the first integral in (D.7) about t_{y+} leads to (Gao, 1999; Gao & Zhong, 1992)

$$\int_{t_1}^{t_2} (\varepsilon - \varepsilon^0) : (D - D^e) : \dot{\varepsilon} dt = (\varepsilon - \varepsilon^0) : [(D - D^e) : \dot{\varepsilon} + \frac{1}{2} \dot{R} : (\varepsilon - \varepsilon^0)] \Delta t$$

$$+ \{\dot{\varepsilon} : (D - D^e) : \dot{\varepsilon} + (\varepsilon - \varepsilon^0) : [(D - D^e) : \ddot{\varepsilon} + \dot{R} : \dot{\varepsilon}]\} \frac{(\Delta t)^2}{2} + O(\Delta t)^3 \tag{D.10}$$

where

$$\dot{R}_{ijkl} = \frac{\partial^2 \hat{\sigma}_{ij}}{\partial \varepsilon_{kl} \partial \varepsilon_{rs}} \dot{\varepsilon}_{rs} - \dot{D}_{ijkl} \tag{D.11}$$

Substituting Eqs. (D.10) and (D.9) into Eq. (D.7), we get

$$(\varepsilon - \varepsilon^0) : \left\{ \dot{G} + (D - D^e) : \dot{\varepsilon} + \frac{1}{2} \dot{R} : (\varepsilon - \varepsilon^0) \right\} \Delta t$$

$$+ \{ \dot{\varepsilon} : [\ddot{G} + (D - D^e) : \dot{\varepsilon}] + (\varepsilon - \varepsilon^0) : [\ddot{G} + (D - D^e) : \ddot{\varepsilon} + \dot{R} : \dot{\varepsilon}] \} \frac{(\Delta t)^2}{2}$$

$$+ O(\Delta t)^3 \leq 0. \tag{D.12}$$

The above inequality is an approximation to the work inequality (D.2). From it, we can derive the flow rule and loading rule. We now divide Eq. (D.12) by Δt, and then let $\Delta t \to 0$, to deduce that

$$(\varepsilon - \varepsilon^0) : [\dot{G} + (D - D^e) : \dot{\varepsilon}] + \frac{1}{2}(\varepsilon - \varepsilon^0) : \dot{R} : (\varepsilon - \varepsilon^0) \leq 0 \tag{D.13}$$

We observe that the first and second terms on the left-hand side are linear and quadratic in $(\varepsilon - \varepsilon^0)$, respectively. Because Eq. (D.13) must hold true for any value of ε^0, we replace $(\varepsilon - \varepsilon^0)$ by $\beta(\varepsilon - \varepsilon^0)$, with β being an arbitrary positive scalar; and then after dividing by β, and then letting $\beta \to 0$, we obtain

$$(\varepsilon - \varepsilon^0) : [\dot{G} + (D - D^e) : \dot{\varepsilon}] \leq 0 \tag{D.14}$$

We now return to inequality (D.12). Again, we replace $(\varepsilon - \varepsilon^0)$ by $\beta(\varepsilon - \varepsilon^0)$ and let $\beta \to 0$. Then, after dividing by Δt and letting $\Delta t \to 0$, we obtain

$$\dot{\varepsilon} : [\dot{G} + (D - D^e) : \dot{\varepsilon}] \leq 0 \tag{D.15}$$

With the help of Eqs. (D.4) and (D.5), inequalities (D.14) and (D.15) can be written as

$$(\varepsilon - \varepsilon^0) : (\dot{\sigma} - D^e : \dot{\varepsilon}) \leq 0 \tag{D.16}$$

and

$$\dot{\varepsilon} : (\dot{\sigma} - D^e : \dot{\varepsilon}) \leq 0 \tag{D.17}$$

We recall that ε^0 is any arbitrary strain inside the loading surface $g(\varepsilon, \varepsilon^p, h^\alpha) = 0$ and note that $\dot{\sigma} - D^e : \dot{\varepsilon}$ is independent of $\varepsilon - \varepsilon^0$ and varies on the loading surface. Therefore, from inequality (D.16) we can conclude that $\dot{\sigma} - D^e : \dot{\varepsilon}$ (which may be thought of as a vector in six-dimensional strain space) must be directed along the normal to the loading surface $g = 0$, that is,

$$\dot{\sigma} - D^e : \dot{\varepsilon} = -\lambda \frac{\partial g}{\partial \varepsilon} \tag{D.18}$$

where λ is a nonnegative scaling factor called the plastic multiplier. Substituting this equation into inequality (D.17) leads to the inequality

$$\frac{\partial g}{\partial \varepsilon} : \dot{\varepsilon} \geq 0 \tag{D.19}$$

Although the quantities with the superposed periods are derivatives with respect to time, they can also be treated as incremental quantities. Thus, we can regard Eq. (D.18) and the inequality (D.19) as incremental relationships. The first of these provides the relationship between the stress increment, strain increment, and the loading function: the so-called plastic flow rule. The inequality (D.19) must hold when plastic strain occurs: this is referred to as the plastic loading rule.

D.2 Derivations for Kinematic Hardening Materials

The yield function for kinematic hardening materials (Eq. 7.59) can be written in the form

$$\bar{f}(\sigma - \rho) - \sigma_y = 0 \tag{D.20}$$

where the parameter ρ is termed the back stress, and the yield stress σ_y is a constant. Differentiating this equation, we obtain

$$
\begin{aligned}
0 &= \frac{\partial \bar{f}}{\partial \sigma} : \dot{\sigma} + \frac{\partial \bar{f}}{\partial \rho} : \dot{\rho} \\
&= \frac{\partial \bar{f}}{\partial \sigma} : \dot{\sigma} - \frac{\partial \bar{f}}{\partial \sigma} : (\sigma - \rho) \frac{\partial \mu}{\partial \bar{\varepsilon}^p} \dot{\bar{\varepsilon}}^p
\end{aligned} \tag{D.21}
$$

From Table 7.1, we observe that $\bar{f}(\sigma)$ is a first-order homogeneous function of stress σ. Hence, using Euler's theorem and Eq. (D.20), we obtain

$$
\begin{aligned}
\frac{\partial \bar{f}}{\partial \sigma} : (\sigma - \rho) &= \bar{f} = \sigma_y \\
\frac{\partial \bar{f}}{\partial \sigma} : \dot{\sigma} &= \dot{\bar{f}}
\end{aligned} \tag{D.22}
$$

Substituting these results into Eq. (D.21), we now obtain

$$
\begin{aligned}
0 &= \dot{\bar{f}} - \sigma_y \frac{\partial \mu}{\partial \bar{\varepsilon}^p} \dot{\bar{\varepsilon}}^p \\
&= \overline{H} \dot{\bar{\varepsilon}}^p - \sigma_y \frac{\partial \mu}{\partial \bar{\varepsilon}^p} \dot{\bar{\varepsilon}}^p
\end{aligned} \tag{D.23}
$$

where the parameter $\overline{H}$ is the slope of the uniaxial stress–plastic strain curve for a kinematic hardening material,

$$\overline{H} = \frac{\partial \bar{f}}{\partial \bar{\varepsilon}^p} \tag{D.24}$$

From Eq. (D.23), we obtain the result quoted in Chapter 7 (Eq. 7.62):

$$\frac{\partial \mu}{\partial \bar{\varepsilon}^p} = \frac{\overline{H}}{\sigma_y} \tag{D.25}$$

To determine an explicit expression for the parameter ψ, we proceed as follows. According to the Ziegler hardening rule (Eq. 7.60), the incremental change in back stress is assumed to be

$$\dot\rho = \dot\mu(\sigma - \rho) \tag{D.26}$$

or, alternatively,

$$\frac{\partial \rho}{\partial \mu} = \sigma - \rho \tag{D.27}$$

Making use of Eq. (D.25), we obtain

$$\frac{\partial \rho}{\partial \bar\varepsilon^p} = \frac{\partial \rho}{\partial \mu}\frac{\partial \mu}{\partial \bar\varepsilon^p} = \frac{\overline{H}}{\sigma_y}(\sigma - \rho) \tag{D.28}$$

Using this result and Eq. (D.20), we use Euler's theorem to obtain the following differentials:

$$\begin{aligned}
\frac{\partial f}{\partial \sigma} &= \frac{\partial \bar f}{\partial \sigma} \\[4pt]
\frac{\partial f}{\partial \varepsilon^p} &= 0 \\[4pt]
\frac{\partial f}{\partial h^\alpha} &= -\frac{\partial \bar f}{\partial \sigma} : \frac{\partial \rho}{\partial \bar\varepsilon^p} \\[4pt]
&= -\frac{\overline{H}}{\sigma_y}\frac{\partial \bar f}{\partial \sigma} : (\sigma - \rho) \\[4pt]
&= -\overline{H}
\end{aligned} \tag{D.29}$$

Substituting these differentials into the general equation (Eq. 7.41), we obtain the explicit form quoted in Chapter 7 (Eq. 7.63):

$$\psi = \frac{\partial \bar f}{\partial \sigma} : D^e : \frac{\partial \bar f}{\partial \sigma} + \overline{H}h_{\varepsilon,\lambda} \tag{D.30}$$

D.3 Derivations for Mixed Hardening Materials

In mixed hardening, the equivalent plastic strain components (isotropic and kinematic) take the role of the hardening parameters, denoted here by h^1 and h^2, respectively. Using the same nomenclature as employed before, we can express these hardening parameters in terms of the plastic multiplier:

$$\begin{aligned}
h^1 &= \dot{\bar\varepsilon}^{pi} = m\dot{\bar\varepsilon}^p = h_{1,\lambda}\dot\lambda \\
h^2 &= \dot{\bar\varepsilon}^{pk} = (1 - m)\dot{\bar\varepsilon}^p = h_{2,\lambda}\dot\lambda
\end{aligned} \tag{D.31}$$

where

$$\begin{aligned}
h_{1,\lambda} &= mh_{\varepsilon,\lambda} \\
h_{2,\lambda} &= (1 - m)h_{\varepsilon,\lambda}
\end{aligned} \tag{D.32}$$

Now, using the results obtained for isotropic hardening (Eq. 7.53) and kinematic hardening (Eq. D.28), we can write

$$\frac{\partial \sigma_y}{\partial \bar{\varepsilon}^{pi}} = H'$$

$$\frac{\partial \rho}{\partial \bar{\varepsilon}^{pk}} = \frac{\overline{H}}{\sigma_y}(\sigma - \rho)$$

(D.33)

Recalling the format of the yield function for mixed hardening (Eq. 7.65),

$$\bar{f}(\sigma - \rho) - \sigma_y(\bar{\varepsilon}^{pi}) = 0$$

(D.34)

we can derive the differentials (making use of Eqs. D.22 and D.33)

$$\frac{\partial f}{\partial \sigma} = \frac{\partial \bar{f}}{\partial \sigma}$$

$$\frac{\partial f}{\partial \varepsilon^p} = 0$$

$$\frac{\partial f}{\partial h^\alpha} h_{\alpha,\lambda} = \frac{\partial f}{\partial h^1} h_{1,\lambda} + \frac{\partial f}{\partial h^2} h_{2,\lambda}$$

$$= -\frac{\partial \sigma_y}{\partial \bar{\varepsilon}^{pi}} h_{1,\lambda} + \frac{\partial \bar{f}}{\partial \rho} : \frac{\partial \rho}{\partial \bar{\varepsilon}^{pk}} h_{2,\lambda}$$

$$= -H' h_{1,\lambda} - \frac{\overline{H}}{\sigma_y} \frac{\partial \bar{f}}{\partial \sigma} : (\sigma - \rho) h_{2,\lambda}$$

$$= -H^* h_{\varepsilon,\lambda}$$

(D.35)

where the parameter H^* is a weighted average of the isotropic and kinematic hardening parameters, that is,

$$H^* = mH' + (1 - m)\overline{H}$$

(D.36)

Substituting these equations into the general equation (Eq. 7.41), we finally obtain the explicit form (Eq. 7.68), as quoted in Chapter 7:

$$\psi = \frac{\partial \bar{f}}{\partial \sigma} : D^e : \frac{\partial \bar{f}}{\partial \sigma} + H^* h_{\varepsilon,\lambda}$$

(D.37)

D.4 Derivation of the Deformation State Function Γ

During the incremental computational process, it is necessary to ascertain at every step whether the material behavior during elasto–plastic loading is hardening or softening. The deformation state parameter Γ performs the function of discriminating between these states. First, we begin by considering the parameter $\hat{f}$, which is defined as

$$\hat{f} = \frac{\partial f}{\partial \sigma} : d\sigma$$

(D.38)

During loading ($f = 0, \hat{g} > 0$), the parameter $\hat{f}$ is positive during hardening, zero during ideal plasticity, and negative during softening. Hence, $\hat{f}$ could be used to discriminate between these states. In particular, because the parameter $\hat{g}$ is always positive during loading, Casey & Naghdi (1981) suggested the use of the ratio $\hat{f}/\hat{g}$ for this purpose, that is,

$$\hat{f}/\hat{g} > 0 \quad \text{(hardening)}$$
$$\hat{f}/\hat{g} = 0 \quad \text{(ideal plasticity)} \tag{D.39}$$
$$\hat{f}/\hat{g} < 0 \quad \text{(softening)}$$

However, since $\hat{f}$ is related to the stress increments, it is not convenient to use it in program code. A better approach can be devised.

We begin by differentiating Eq. (7.23), which yields

$$\hat{f} + \frac{\partial f}{\partial \varepsilon^p} : \dot{\varepsilon}^p + \frac{\partial f}{\partial h^\alpha} h^\alpha = \hat{g} + \frac{\partial g}{\partial \varepsilon^p} : \dot{\varepsilon}^p + \frac{\partial g}{\partial h^\alpha} h^\alpha \tag{D.40}$$

Using Eqs. (7.32) and (7.36), we can write this equation as

$$\hat{f} = \hat{g} + \lambda \left(\frac{\partial g}{\partial \varepsilon^p} - \frac{\partial f}{\partial \varepsilon^p} \right) : \frac{\partial f}{\partial \sigma} + \lambda \left(\frac{\partial g}{\partial h^\alpha} - \frac{\partial f}{\partial h^\alpha} \right) h_{\alpha,\lambda} \tag{D.41}$$

Then, using Eqs. (7.27) and (7.39), we obtain

$$\hat{f} = \hat{g} - \lambda \frac{\partial f}{\partial \sigma} : \mathbf{D}^e : \frac{\partial f}{\partial \sigma} \tag{D.42}$$

Substituting Eqs. (7.40) and (7.43) into this equation leads to

$$\hat{f} = \hat{g} - \frac{\hat{g}}{\psi} \frac{\partial f}{\partial \sigma} : \mathbf{D}^e : \frac{\partial f}{\partial \sigma} \tag{D.43}$$

Finally, substituting Eq. (7.41) into this equation, we obtain

$$\hat{f}/\hat{g} = \Gamma/\psi \tag{D.44}$$

where

$$\Gamma = -\frac{\partial f}{\partial \varepsilon^p} : \frac{\partial f}{\partial \sigma} - \frac{\partial f}{\partial h^\alpha} h_{\alpha,\lambda} \tag{D.45}$$

From Eqs. (7.26) and (7.37), we can see that, during loading, the parameter ψ is nonnegative. Hence, the discriminating function (D.39) can be replaced by

$$\Gamma > 0 \quad \text{(hardening)}$$
$$\Gamma = 0 \quad \text{(ideal plasticity)} \tag{D.46}$$
$$\Gamma < 0 \quad \text{(softening)}$$

From Eq. (D.45), it is evident that the parameter Γ is independent of stress (or strain) rates. It is, instead, a function of the stress (or strain) state. In terms of programming, this turns out to be a much more convenient measure of the deformation state.

APPENDIX E

Domain Integral Formulations

E.1 Boundary Integral Equations: Initial Strain Formulation

The "initial stress" approach is preferred in this book. However, it is quite possible to recast the equations in terms of "initial strains," if desired. First, we note that initial strain is defined as the irrecoverable strain, which, in the elasto–plastic context, is the plastic strain. Substituting the stress–strain equation (Eq. 8.5) into the boundary integral equation (Eq. 8.14), we obtain the elasto–plastic boundary integral equations, in terms of initial strains:

$$c_{ij}(P)\dot{u}_j(P) + \int_\Gamma T_{ij}(P, Q)\dot{u}_j(Q)d\Gamma(Q) = \int_\Gamma U_{ij}(P, Q)\dot{t}_j(Q)d\Gamma(Q)$$

$$+ \int_\Omega U_{ij}(P, q)b_j(q)d\Omega(q) + \int_\Omega \Sigma_{ijk}(P, q)\dot{\varepsilon}_{jk}^p(q)d\Omega(q), \qquad \text{(E.1)}$$

where

$$\Sigma_{ijk}(P, q) = D_{jkrs}^e E_{irs}(P, q) \qquad \text{(E.2)}$$

E.2 Analytical Integration of the Strongly Singular Volume Integral

For the three-dimensional case, we define a spherical coordinate system with origin at p, as shown in Fig. E.1. The relationships between the Cartesian and spherical systems are

$$x = r \sin\theta \cos\phi$$
$$y = r \sin\theta \sin\phi \qquad \text{(E.3)}$$
$$z = r \cos\theta \qquad 0 \le \theta \le \pi, \ 0 \le \phi \le 2\pi$$

$$d\Omega = dSdr \qquad \text{(E.4)}$$

where dS is a differential element on a spherical surface, that is,

$$dS = r^2 \sin\theta d\theta d\phi \qquad \text{(E.5)}$$

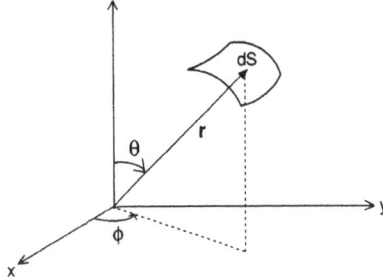

Figure E.1: Spherical coordinate system.

The term $r_{,i}$ can be written as

$$r_{,i} = \frac{\partial r}{\partial x_i} = \frac{\partial r}{\partial x}\frac{\partial x}{\partial x_i} + \frac{\partial r}{\partial y}\frac{\partial y}{\partial x_i} + \frac{\partial r}{\partial z}\frac{\partial z}{\partial x_i} \tag{E.6}$$

$$= \delta_{1i} \sin\theta \cos\phi + \delta_{2i} \sin\theta \sin\phi + \delta_{3i} \cos\theta$$

We can easily show that

$$\int_0^\pi \int_0^{2\pi} \sin\theta \, d\phi \, d\theta = 4\pi \tag{E.7}$$

Using Eq. (E.6) and noting that $\delta_{1i}\delta_{1j} + \delta_{2i}\delta_{2j} + \delta_{3i}\delta_{3j} = \delta_{ij}$, we obtain

$$\int_0^\pi \int_0^{2\pi} r_{,i}\, r_{,j} \sin\theta \, d\phi \, d\theta = \frac{4\pi}{3}\delta_{ij} \tag{E.8}$$

and

$$\int_0^\pi \int_0^{2\pi} r_{,i}\, r_{,j}\, r_{,k}\, r_{,l} \sin\theta \, d\phi \, d\theta = \frac{4\pi}{15}(\delta_{ij}\delta_{kl} + \delta_{ik}\delta_{jl} + \delta_{il}\delta_{jk}) \tag{E.9}$$

Combining these results (here, $d\Gamma = dS$) and noting that $r = \varepsilon$ and $n_m = \partial r/\partial x_m = r_{,m}$ we obtain the equation quoted in Chapter 8 (Eq. 8.19):

$$\int_{\Omega_\varepsilon} \frac{\partial E_{ijk}(p,q)}{\partial x_m^p} d\Omega(q) = \frac{-1}{30(1-v)G}\{\delta_{im}\delta_{jk} - (4-5v)(\delta_{ij}\delta_{km} + \delta_{ik}\delta_{jm})\} \tag{E.10}$$

Similarly, for the two-dimensional case, using a polar coordinate system and the relationships

$$x = r \cos\theta$$
$$y = r \sin\theta \tag{E.11}$$
$$d\Gamma = r\, d\theta$$

and

$$r_{,i} = \frac{\partial r}{\partial x_i}$$

$$= \frac{\partial r}{\partial x}\frac{\partial x}{\partial x_i} + \frac{\partial r}{\partial y}\frac{\partial y}{\partial x_i} \tag{E.12}$$

$$= \delta_{1i} \cos\theta + \delta_{2i} \sin\theta$$

we can obtain

$$\int_0^{2\pi} r_{,i}\, r_{,j}\, d\theta = \pi \delta_{ij} \tag{E.13}$$

$$\int_0^{2\pi} r_{,i}\, r_{,j}\, r_{,k}\, r_{,l}\, d\theta = \frac{\pi}{4}(\delta_{ij}\delta_{kl} + \delta_{ik}\delta_{jl} + \delta_{il}\delta_{jk}) \tag{E.14}$$

Combining these results, we obtain the equation quoted in Chapter 8 (Eq. 8.20):

$$\int_{\Omega_\varepsilon} \frac{\partial E_{ijk}(p,q)}{\partial x_m^p} d\Omega(q) = \frac{-1}{16(1-\nu)G}\{\delta_{im}\delta_{jk} - (3-4\nu)(\delta_{ij}\delta_{km} + \delta_{ik}\delta_{jm})\} \tag{E.15}$$

E.3 Interior Stress Equation: Initial Strain Formulation

In an initial strain formulation, Eq. (8.21) can be transformed into its initial strain equivalent, by making use of the stress–strain equation (Eq. 8.5), to obtain

$$\dot{\sigma}_{ij}(p) = \int_\Gamma U_{ijk}(p,Q)\dot{t}_k(Q)d\Gamma(Q) - \int_\Gamma T_{ijk}(p,Q)\dot{u}_k(Q)d\Gamma(Q)$$

$$+ \int_\Omega U_{ijk}(p,q)b_k(q)d\Omega(q) + \int_\Omega \Sigma_{ijkl}(p,q)\dot{\varepsilon}_{kl}^p(q)d\Omega(q) \tag{E.16}$$

$$+ F_{ijkl}^\varepsilon \dot{\varepsilon}_{kl}^p(p)$$

where

$$\Sigma_{ijkl} = E_{ijrs}D_{rskl}^e \tag{E.17}$$

$$F_{ijkl}^\varepsilon = F_{ijrs}^\sigma D_{rskl}^e \tag{E.18}$$

These equations are equivalent to the initial stress equations employed in the text.

E.4 Analytical Integration of E_{ijkl} in Two Dimensions

The procedure employed to integrate the function E_{ijkl} over a symmetric region around the singularity is precisely the same whether one considers the three-dimensional case or the two-dimensional case. The first of these is described in some detail in Chapter Eight, but a brief summary of the latter is presented here for completeness. In two dimensions, the sphere of exclusion reduces to a circle centered on the singular point p. Using a polar coordinate (r,θ) system, as depicted in Fig. E.2, we obtain, using Eqs. (E.11) and Eq. (8.29),

$$\int_{\Omega_s} E_{ijkl}d\Omega = \int_0^{2\pi} \left[\lim_{\varepsilon\to 0}\int_\varepsilon^{r(\Gamma_s)}\frac{1}{r}dr\right]\Psi_{ijkl}d\theta$$

$$= \int_0^{2\pi}\Psi_{ijkl}\log_e r(\Gamma_s)d\theta - \lim_{\varepsilon\to 0}\log_e \varepsilon\int_0^{2\pi}\Psi_{ijkl}d\theta \tag{E.19}$$

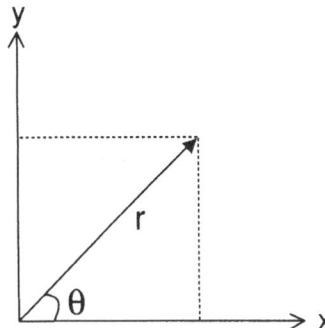

Figure E.2: Polar coordinate system.

Now, using Eqs. (E.13) and (E.14), we can prove that

$$\int_0^{2\pi} \Psi_{ijkl} d\theta = 0 \tag{E.20}$$

Furthermore, from Fig. E.3, it is clear that

$$r d\theta = d\Gamma \cos\varphi$$
$$= d\Gamma \frac{r_i n_i}{r} \tag{E.21}$$

where φ is the angle between the normal to the boundary and the radial vector from the singular point.

Substituting Eqs. (E.20) and (E.21) into Eq. (E.19), we obtain

$$\int_{\Omega_s} E_{ijkl} d\Omega = \int_{\Gamma_s} E_{ijkl} r_m n_m \log_e r \, d\Gamma \tag{E.22}$$

As can be seen, this equation has precisely the same form as the three-dimensional equivalent (Eq. 8.35).

E.5 Analytical Integration: Initial Strain Formulation

In the initial stress formulation, a singular volume integral is reduced to a surface integral by analytical means. This transformation, captured by Eq. (8.35), takes the following equivalent form in an initial strain formulation:

$$\int_{\Omega_s} \Sigma_{ijkl} d\Omega = \int_{\Gamma_s} \Sigma_{ijkl} r_m n_m \log_e r \, d\Gamma \tag{E.23}$$

where we have made use of Eq. (E.17).

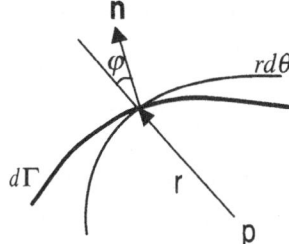

Figure E.3: Relation between circular and cell boundaries.

Solution of the Nonlinear System Equations

F.1 The Newton–Raphson Iterative Algorithm

The Newton–Raphson method (also called Newton's method) is a popular method for solving sets of nonlinear equations. We assume that there are N equations, which assume the functional form

$$f_i(x_1, x_2, \ldots, x_N) = 0 \tag{F.1}$$

where x_j are the variables. Beginning with some initial values for these variables, more accurate values are then obtained by iteration. We denote the values of the variables after the kth iteration by the symbol x_j^k. If these values are inaccurate, the functions f_i are not, in general, satisfied exactly. We denote the residuals of these functions by the symbols f_i^k, that is,

$$f_i^k = f_i(x_1^k, x_2^k, \ldots, x_N^k) \tag{F.2}$$

To reduce these residuals to zero, the values of the variables must be corrected. Denoting the corrections to these values by Δx_j, we write the updated values (for the next iteration) as

$$x_j^{k+1} = x_j^k + \Delta x_j \tag{F.3}$$

To determine the changes Δx_j, we expand the functions f_i using the Taylor series

$$f_i^{k+1} = f_i^k + \frac{\partial f_i}{\partial x_j} \Delta x_j + O(\Delta x_j^2) \tag{F.4}$$

Neglecting the higher-order terms and forcing $f_i^{k+1} = 0$ yields the set of linear equations

$$\frac{\partial f_i}{\partial x_j} \Delta x_j = -f_i^k \tag{F.5}$$

Once this set of equations has been solved for the corrections Δx_j, the variables are updated using Eq. (F.3) and the process is repeated until satisfactory convergence

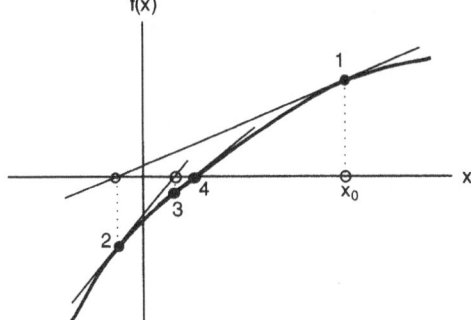

Figure F.1: The Newton–Raphson process.

is attained. To illustrate the algorithm, Fig. F.1 depicts the progress of the algorithm for the case of a single variable x, where x_0 is the initial guess.

Since the Newton–Raphson method makes use of the gradient of the functions to update the solution, the process is rapidly convergent, provided that the initial guess is not too far away from the correct solution. However, if the initial guess is poor, then the process may not converge at all, or worse still, it may converge to the wrong solution. In the incremental solution of the nonlinear equations of solid mechanics, failure of convergence is not observed in general, except where it reflects physical instabilities.

F.2 System Equation Solution Strategies

Aside from the method of solution described in Chapter Nine, the existing strategies for the solution of the nonlinear boundary element system equations can be broadly categorized into four groups:

(i) the initial stress (or strain) iteration technique (Telles & Brebbia, 1979; Banerjee & Davies, 1979; Telles, 1983);
(ii) the implicit solution technique (Telles & Carrer, 1991; Bonnet & Mukherjee, 1996);
(iii) the variable stiffness technique (Banerjee et al., 1989; Banerjee, 1994); and
(iv) the mixed representation technique (Chopra & Dargush, 1994).

Each of these is briefly described in turn below, but for convenience we first gather together the pertinent equations from earlier chapters. From Chapter Seven, we recall Eqs. (7.32) and (7.40):

$$\dot{\varepsilon}^p = \lambda \frac{\partial f}{\partial \sigma} \tag{F.6}$$

$$\lambda = \frac{1}{\psi} \frac{\partial f}{\partial \sigma} : \mathbf{D}^e : \dot{\varepsilon} \tag{F.7}$$

From Chapter Eight, we recall Eqs. (8.3), (8.4), and (8.5):

$$\dot{\sigma}_{ij} = \dot{\sigma}_{ij}^e - \dot{\sigma}_{ij}^p \tag{F.8}$$

$$\dot{\sigma}_{ij}^e = D_{ijkl}^e \dot{\varepsilon}_{kl} \tag{F.9}$$

$$\dot{\sigma}_{ij}^p = D_{ijkl}^e \dot{\varepsilon}_{kl}^p \tag{F.10}$$

And, from Chapter Nine, we recall Eqs. (9.22), (9.24), (9.25), and (9.28):

$$[A^b]\{\dot{x}\} = \{\dot{y}^b\} + [E^b]\{\dot{\sigma}^P\} \tag{F.11}$$

$$\{\dot{\sigma}\} = [A^\sigma]\{\dot{x}\} + \{\dot{y}^\sigma\} + [E^\sigma]\{\dot{\sigma}^P\} \tag{F.12}$$

$$\{\dot{u}\} = [A^u]\{\dot{x}\} + \{\dot{y}^u\} + [E^u]\{\dot{\sigma}^P\} \tag{F.13}$$

$$\{d^f\} = [D^e]\left\{\frac{\partial f}{\partial \sigma}\right\} \tag{F.14}$$

Discussion of these equations is omitted here in order to avoid unnecessary repetition.

F.2.1 The Initial Stress Iteration Technique

From Eqs. (F.6), (F.7), and (F.10), we can express the initial stress increments in the form

$$\{\dot{\sigma}^P\} = [C^p]\{\dot{\sigma}^e\} \tag{F.15}$$

where

$$[C^p] = \frac{1}{\psi}[D^e]\left\{\frac{\partial f}{\partial \sigma}\right\}\left\{\frac{\partial f}{\partial \sigma}\right\}^T \tag{F.16}$$

Using Eq. (F.15), we can rewrite Eq. (F.11) as:

$$[A^b]\{\dot{x}\} = \{\dot{y}^b\} + [E^b][C^p]\{\dot{\sigma}^e\} \tag{F.17}$$

and, from Eq. (F.12), the stresses can be expressed in terms of the elastic stresses and so

$$\{\dot{\sigma}\} = [A^\sigma]\{\dot{x}\} + \{\dot{y}^\sigma\} + [E^\sigma][C^p]\{\dot{\sigma}^e\} \tag{F.18}$$

After each iteration, the new elastic stresses can be determined from Eqs. (F.8) and (F.15), that is,

$$\{\dot{\sigma}^e\} = \{\dot{\sigma}\} + \{\dot{\sigma}^P\} \tag{F.19}$$

The incremental iteration scheme can be outlined as follows:

1. Obtain an elastic solution and scale to first yield. Determine $\{\dot{y}^b\}$ and $\{\dot{y}^\sigma\}$ for every load increment. Set all elastic stress increments $\{\dot{\sigma}^e\}$ to zero.
2. Apply load increments $\{\dot{y}^b\}$ and $\{\dot{y}^\sigma\}$ for the first iteration.
3. Solve Eq. (F.17) for boundary unknowns $\{\dot{x}\}$ ($\{\dot{y}^b\} = 0$ after first iteration).
4. Use the solved $\{\dot{x}\}$ to calculate stress $\{\dot{\sigma}\}$ from Eq. (F.18) ($\{\dot{y}^\sigma\} = 0$ after first iteration).

5. If desired, calculate displacements using Eqs. (F.13) and (F.15) similarly.
6. Update the stresses and boundary unknowns (and displacements).
7. Calculate the new elastic stress $\{\dot{\sigma}^e\}$ using Eqs. (F.19) and (F.15).
8. If the norm of the vector $\{\dot{\sigma}^e\}$ is sufficiently small, go to step 2. Otherwise, go to step 3 for next iteration.

In this technique, the boundary unknowns are the primary variables of the system equations and hence the computer memory requirement is modest. However, the method suffers from very slow convergence and, in some difficult cases, convergence may be not achieved. The initial strain technique is very similar to this and need not be discussed separately.

F.2.2 The Implicit Solution Technique

In the implicit solution technique, the final system equations are expressed in terms of elastic stress (or total strain) increments. This can be done by eliminating the boundary unknowns from Eqs. (F.17) and (F.18). This results in the following expression:

$$[C^{\sigma}]\{\dot{\sigma}^e\} = \{\dot{y}^e\} \tag{F.20}$$

where

$$\{\dot{y}^e\} = \{\dot{y}^{\tau}\} + [A^{\tau}][A^b]^{-1}\{\dot{y}^b\} \tag{F.21}$$

and

$$[C^{\sigma}] = [I] - ([I] + [E^{\sigma}] + [A^{\sigma}][A^b]^{-1}[E^b])[C^p] \tag{F.22}$$

in which $[I]$ is the identity matrix. Equation (F.20) is an equation set in terms of elastic stress increments. Alternatively, substituting Eq. (F.9) into it, we can obtain the equations in terms of the true strain increments as follows:

$$[K]\{\dot{\varepsilon}\} = \{\dot{y}^e\} \tag{F.23}$$

where

$$[K] = [C^{\sigma}][D^e] \tag{F.24}$$

These equations can be solved using various incremental iterative schemes (e.g., Gao, 1999). In each iteration, because the stresses (or strains) are directly solved from the system equations, convergence is rapid. For example, Bonnet & Mukherjee (1996) applied the consistent tangent operator method (proposed by Simo & Taylor (1985) in the finite element method context) to Eq. (F.23). Since the second derivatives of yield function with respect to the stress in the constitutive relationship are used in this method, a quadratic rate of convergence can be achieved in the Newton–Raphson iterative process. Although the implicit method is relatively easy to code and is rapidly convergent, it imposes considerable computer memory requirements. This is because the system equations are formulated in terms of stress (or strain) increments, which have six degrees of freedom at each node, in three dimensions. Even if the block decomposition technique is employed, as described by Bonnet & Mukherjee (1996), the computer memory requirement is substantial.

F.2.3 The Variable Stiffness Technique

The variable stiffness technique was proposed by Banerjee et al. (1989) and has been successfully demonstrated by Banerjee and his co-workers (Raveendra, 1984; Banerjee, 1994). From Eqs. (F.6) and (F.10), the initial stress can be expressed in the form

$$\{\dot{\sigma}^p\} = [d^f]\{\dot{\lambda}\} \tag{F.25}$$

where $[d^f]$ is a diagonally dominant sparse matrix. Substituting this equation into Eqs. (F.11) and (F.12) yields

$$[A^b]\{\dot{x}\} = \{\dot{y}^b\} + [E^b][d^f]\{\dot{\lambda}\} \tag{F.26}$$

and

$$\{\dot{\sigma}\} = [A^\sigma]\{\dot{x}\} + \{\dot{y}^\sigma\} + [E^\sigma][d^f]\{\dot{\lambda}\} \tag{F.27}$$

From Eqs. (F.7) and (F.9), the plastic multiplier at a node can be written in the form

$$\{\dot{\lambda}\} = [\nabla f_\psi]\{\dot{\sigma}^e\} \tag{F.28}$$

where

$$[\nabla f_\psi] = \frac{1}{\psi}\left\{\frac{\partial f}{\partial \sigma}\right\}^T \tag{F.29}$$

Now, substituting Eq. (F.25) into Eq. (F.19) and putting the result, together with Eq. (F.27), into Eq. (F.28) yields

$$[H^\lambda]\{\dot{\lambda}\} = [A^x]\{\dot{x}\} + \{\dot{y}^\lambda\} \tag{F.30}$$

where

$$[H^\lambda] = [I] - [\nabla f_\psi]([I] + [E^\sigma])[d^f] \tag{F.31}$$

$$[A^x] = [\nabla f_\psi][A^\sigma] \tag{F.32}$$

and

$$\{\dot{y}^\lambda\} = [\nabla f_\psi]\{\dot{y}^\sigma\} \tag{F.33}$$

Inverting Eq. (F.30) produces

$$\{\dot{\lambda}\} = [H^\lambda]^{-1}[A^x]\{\dot{x}\} + [H^\lambda]^{-1}\{\dot{y}^\lambda\} \tag{F.34}$$

Finally, substituting this equation into Eq. (F.26), we obtain

$$[A]\{\dot{x}\} = \{\dot{y}\} \tag{F.35}$$

where

$$[A] = [A^b] - [E^b][d^f][H^\lambda]^{-1}[A^x] \tag{F.36}$$

and

$$\{\dot{y}\} = \{\dot{y}^b\} + [E^b][d^f][H^\lambda]^{-1}\{\dot{y}^\lambda\} \tag{F.37}$$

Equation (F.35) is constructed and solved for the boundary unknowns $\{\dot{x}\}$ for each load increment. Once these unknowns are obtained, one can obtain $\{\dot{\lambda}\}$ from Eq. (F.34). Then, the stress increment $\{\dot{\sigma}\}$ can be computed using Eq. (F.27). In this variable stiffness solution scheme (described in detail by Banerjee, 1994), the internal variables (plastic multipliers) are eliminated, by expressing them in terms of boundary variables, and consequently no iteration is needed, provided that small increments are used. Another advantage is that the primary variables are the boundary unknowns only. However, its principal disadvantage is that the solution may drift, as no iteration process is employed to control it.

F.2.4 The Mixed Representation Technique

Rather than eliminating the plastic multipliers, Chopra & Dargush (1994) took both the boundary unknowns and the plastic multipliers to be the primary unknowns in the system equations. They then solved the system equations using the Newton–Raphson method. The solution begins with the observation that Eqs. (F.26) and (F.30) are not, in general, satisfied exactly during an incremental process. The residuals of these two sets of equations are defined as

$$\{R_b\} = [A^b]\{\dot{x}\} - [E^b][d^f]\{\dot{\lambda}\} - \{\dot{y}^b\} \tag{F.38a}$$

$$\{R_\lambda\} = [A^x]\{\dot{x}\} - [H^\lambda]\{\dot{\lambda}\} + \{\dot{y}^\lambda\} \tag{F.38b}$$

Following the Newton–Raphson strategy, the Taylor series expansions of these residuals, after the kth iteration, are

$$\{R_b^{k+1}\} = \{R_b^k\} + \frac{\partial\{R_b\}}{\partial\{\dot{x}\}}\{\Delta\dot{x}\} + \frac{\partial\{R_b\}}{\partial\{\dot{\lambda}\}}\{\Delta\dot{\lambda}\} + O(\Delta\dot{x}^2, \Delta\dot{\lambda}^2) \tag{F.39a}$$

$$\{R_\lambda^{k+1}\} = \{R_\lambda^k\} + \frac{\partial\{R_\lambda\}}{\partial\{\dot{x}\}}\{\Delta\dot{x}\} + \frac{\partial\{R_\lambda\}}{\partial\{\dot{\lambda}\}}\{\Delta\dot{\lambda}\} + O(\Delta\dot{x}^2, \Delta\dot{\lambda}^2) \tag{F.39b}$$

where $\{R_b^k\}$ and $\{R_\lambda^k\}$ are the residuals calculated from Eq. (F.38) using quantities obtained at the end of the kth iteration, that is,

$$\{R_b^k\} = [A^b]\{\dot{x}^k\} - [E^b][d^f]\{\dot{\lambda}^k\} - \{\dot{y}^b\} \tag{F.40a}$$

$$\{R_\lambda^k\} = [A^x]\{\dot{x}^k\} - [H^\lambda]\{\dot{\lambda}^k\} + \{\dot{y}^\lambda\} \tag{F.40b}$$

The derivatives in Eqs. (F.39) can be readily obtained from Eqs. (F.38), and so

$$
\frac{\partial \{R_b\}}{\partial \{\dot{x}\}} = [A^b]
$$

$$
\frac{\partial \{R_b\}}{\partial \{\lambda^k\}} = -[E^b][d^f]
$$

$$
\frac{\partial \{R_{\lambda_\lambda}\}}{\partial \{\dot{x}\}} = [A^x] \tag{F.41}
$$

$$
\frac{\partial \{R_\lambda\}}{\partial \{\lambda^k\}} = -[H^\lambda]
$$

Neglecting the higher-order terms in Eqs. (F.39) and forcing the residuals $\{R_b^{k+1}\}$ and $\{R_\lambda^{k+1}\}$ to be zero, we obtain the following coupled set of equations:

$$
\begin{bmatrix} [A^b] & -[E^b][d^f] \\ [A^x] & -[H^\lambda] \end{bmatrix} \begin{Bmatrix} \{\Delta\dot{x}\} \\ \{\Delta\lambda\} \end{Bmatrix} = \begin{Bmatrix} -\{R_b^k\} \\ -\{R_\lambda^k\} \end{Bmatrix} \tag{F.42}
$$

After solving this set of equations for $\{\Delta\dot{x}\}$ and $\{\Delta\lambda\}$, the boundary unknowns and plastic multipliers are updated, using the equations

$$
\{\dot{x}^{k+1}\} = \{\dot{x}^k\} + \{\Delta\dot{x}\}
$$

$$
\{\lambda^{k+1}\} = \{\lambda^k\} + \{\Delta\lambda\} \tag{F.43}
$$

Further details of this iterative process are given by Chopra & Dargush (1994). This method should converge quickly but, because both boundary unknowns and plastic multipliers are primary variables, the computer memory requirement is substantial.

APPENDIX G

Elements of Elasto–Plasticity

The development of elasto–plastic theory in Chapter Seven and the derivations in Appendix D are based on the strain-space approach. Here, we develop much the same equations, using the simpler stress-space approach. Strictly speaking, they are not applicable to strain-softening materials, that is, those where the yield stress decreases with increasing plastic strain. For the benefit of those new to elasto–plastic flow theory, the equations are derived from simple physical arguments wherever possible.

A basic premise of elasto–plastic flow theory is that plastic deformations are "path independent in small," that is, the direction in which the material flows (the resultant vector of the plastic strain increments) depends on the current state of the material but not on the stress increment itself. In that respect, plastic flow shares some similarity to potential flows (heat, electricity, etc.). The state of stress in a material undergoing plastic flow must satisfy the yield function

$$f(\sigma) = 0 \qquad (G.1)$$

because any stress state "higher" than yield is inadmissible, whereas any stress state "lower" than yield will be elastic. The yield function can be thought of as a surface in a multidimensional space (stress space) in which the axes are the stress components. Since during plastic loading, the material moves from one plastic state to another, it follows that during this excursion, the total derivative of the yield function must be zero, that is,

$$df = 0 \qquad (G.2)$$

This is known as the "consistency" condition. During elasto–plastic deformation, we assume that the total strain increment can be decomposed into elastic (recoverable) and plastic (irrecoverable) parts as follows:

$$\dot{\varepsilon} = \dot{\varepsilon}^e + \dot{\varepsilon}^p \qquad (G.3)$$

where the superposed period indicates an incremental quantity.

According to Drucker's (1959) postulate, the work done during a closed stress cycle must be nonnegative. The work done in an elemental volume of material during an increment of stress (that is, $\dot{\sigma}$) can be calculated from the scalar product of the stress and strain increment vectors. Over a closed stress cycle, the elastic strain increment is recovered, and thus the net work done is related to the plastic strain increment only. Drucker's postulate then results in the inequality

$$\dot{\sigma} : \dot{\varepsilon}^p \geq 0 \qquad\qquad\qquad (G.4)$$

where the colon indicates the scalar product of the two vectors. In vector analysis, the scalar product of two vectors is greater than zero only if the angle between the two vectors is less than 90°. Now, invoking the assumption of "path independence in small," the plastic strain increment vector is a function of the current stress state and not the stress increment. Because the stress increment vector can be directed outward from the yield surface in any direction, the only direction in which the plastic strain increment vector can be oriented is normal to the surface. Only if it is oriented in this direction is the angle between the two vectors always less than 90°. This result, known as the "normality" principle, can be expressed in the form

$$\dot{\varepsilon}^p = \lambda \frac{\partial f}{\partial \sigma} \qquad\qquad\qquad (G.5)$$

where λ is a nonnegative parameter, called the plastic multiplier.

This equation, called the flow rule, can be combined with the consistency condition and the elastic stress–strain relationship to derive the elasto–plastic constitutive relationships. Using the elastic stress–strain relationships and the decomposition of incremental strain into recoverable and irrecoverable components (Eq. G.3), we obtain

$$\begin{aligned}\dot{\sigma} &= \mathbf{D}^e : \dot{\varepsilon}^e \\ &= \mathbf{D}^e : (\dot{\varepsilon} - \dot{\varepsilon}^p)\end{aligned} \qquad\qquad\qquad (G.6)$$

where $\mathbf{D}^e$ is the elastic constitutive tensor. For a specific material, the yield function can be written as

$$f(\sigma, k) = 0 \qquad\qquad\qquad (G.7)$$

where k is some material (hardening) parameter, which is normally a function of the accumulated plastic strain ε^p. Using the consistency condition, it follows that

$$\frac{\partial f}{\partial \sigma} : \dot{\sigma} + \frac{\partial f}{\partial k}\frac{\partial k}{\partial \varepsilon^p} : \dot{\varepsilon}^p = 0 \qquad\qquad\qquad (G.8)$$

Substituting Eqs. (G.5) and (G.6) into this equation, and solving for the plastic multiplier, we obtain

$$\lambda = \frac{1}{\psi}\frac{\partial f}{\partial \sigma} : \mathbf{D}^e : \dot{\varepsilon} \qquad\qquad\qquad (G.9)$$

where the parameter ψ is

$$\psi = \frac{\partial f}{\partial \sigma} : \mathbf{D}^e : \frac{\partial f}{\partial \sigma} - \frac{\partial f}{\partial k}\frac{\partial k}{\partial \varepsilon^p} : \frac{\partial f}{\partial \sigma} \qquad\qquad\qquad (G.10)$$

Substituting Eqs. (G.5), (G.9), and (G.10) into Eq. (G.6), we obtain the incremental elasto–plastic stress–strain relationship

$$\dot{\sigma} = \left(\mathbf{D}^e - \frac{1}{\psi} \mathbf{D}^e : \frac{\partial f}{\partial \sigma} \otimes \frac{\partial f}{\partial \sigma} : \mathbf{D}^e \right) : \dot{\varepsilon} \qquad (G.11)$$

where the symbol $\otimes$ indicates the "outer product," that is, the matrix resulting from the product of the vector $(\partial f/\partial \sigma)$ and its transpose. To apply this equation in practice, experimental data are needed to define the yield function f and how it evolves with respect to the cumulative plastic strain.

Description of Input Data

The input data for program BEMECH is organized into fifteen "card sets," invoking memories of the time when data were entered on Hollerith punched cards. Thus, a set of data that should appear on a single line in the input file is termed a "card" here. All data are entered in free format. In addition to the notes here, more detailed descriptions of the input variables can be found in Chapters Five and Ten, while some simple data sets are reproduced in Chapters Six and Eleven. The superscripted numbers in the following refer to the Notes at the end of this appendix.

Card Set 1 (one card)
TITLE: Alpha-numerical characters that identify the problem.

Card Set 2 (one card)
NSIG: Number of stress components (3, 4, or 6).[1]
NODE: Number of boundary element nodes (2, 3, 4, or 8).[2]
NBP: Total number of boundary nodes.[3]
NTP: Total number of boundary and interior nodes.[4]
NBE: Total number of boundary elements.
NCELL: Total number of interior cells. Input *zero* for elasticity.
NTRAC: Number of specified traction groups.[5]
NDISP: Number of specified displacement groups.[6]
KSYM: Symmetry condition flag (0–7).[7]
NAUTO: Number of Gauss points used to evaluate integrals.[8]

Card Set 3 (one card)
TOLGP: Tolerance for automatic Gauss integration.[9]
CSYM(1):   Coordinate x_c of symmetry plane. *Zero*, if none.
CSYM(2):   Coordinate y_c of symmetry plane. *Zero*, if none.
CSYM(3):   Coordinate z_c of symmetry plane (omit in two dimensions).

Card Set 4 (NTP cards – one for each node)
I: Node number.[10]

CD(1,I): x-coordinate of node I.
CD(2,I): y-coordinate of node I.
CD(3,I): z-coordinate of node I (omit in two dimensions).

Card Set 5 (one card)
NFIXU: Number of constrained displacement nodes.[11]

Card Set 6 (NFIXU cards – one for each constrained node)
J: Constrained node number.
NUGRP(J):   Group number of prescribed displacements for node J.[12]
KODP(J):   Flags for constrained displacement directions.[13]

Card Set 7 (one card)
MULTP: Number of traction-discontinuous nodes.[14]

Card Set 8 (MULTP cards – one for each traction-discontinuous node)
K: Additional node number for traction-discontinuous node.[15]
KBU(K):   Corresponding basic boundary node number.[16]

Card Set 9 (NBE cards – one for each boundary element)
L: Boundary element number.
LNDB(L,1):   Global node number for the 1st node of element L.[17]
LNDB(L,2):   Global node number for the 2nd node . . . .
LNDB(L,ω):   Global node number for the last node of element L.
NTGRP(L):   Group numbers of prescribed tractions for element L.
IFLAG: Flag for prescribed traction directions.[18]

Card Set 10 (NTRAC cards – one for each prescribed traction group)
M: Group number.
F: Array containing the specified tractions for all element nodes.[19]

Card Set 11 (NDISP cards – one for each prescribed displacement group)
N: Group number.
RU: Array containing specified displacements at the node.[20]

Card Set 12 (one card)
DIAGV: Problem type (closed, semi-infinite, or infinite region).[21]
E: Young's modulus of elasticity.
PR: Poisson's ratio.

Card Set 13 (NCELL cards – one card for each cell; omit in elastic analysis)
P: Cell number.
LNDC(1,P):   Global node number of 1st node of cell P.[17]
LNDC(2,P):   Global node number of 2nd node . . . .
LNDC(ω,P):   Global node number of last node of cell P.

Card Set 14 (one card; omit in elastic analysis)
NCRIT: Flag for yield criterion (1, 2, 3, or 4).[22]
UNIAX: Uniaxial initial yield stress σ_y or cohesion c.[23]
FRICT: Friction angle ϕ (degrees).[24]

HARDI: Isotropic hardening modulus $H.'^{25}$
HARDK: Kinematic hardening modulus $\overline{H}.^{25}$
EM: Scale factor for mixed hardening $(0-1).^{26}$

Card Set 15 (one card; omit in elastic analysis)
NINCS: Number of load increments.[27]
FINCR: Ratio of two successive load increments, φ^{28}.
MITER: Maximum number of iterations permitted per increment.[29]
TOLER: Convergence tolerance during iteration.[30]
NOUT1:   Output flag, for each increment.[31]
NOUT2:   Output flag, for final results.[32]

NOTES
(1) NSIG: 3 = plane stress, 4 = plane strain, 6 = three-dimensional.

(2) NODE: 2 = 2D (linear), 3 = 2D (quadratic), 4 = 3D (linear), 8 = 3D (quadratic).

(3) NBP: Excluding any additional "traction-discontinuous" nodes.

(4) NTP: Including any additional "traction-discontinuous" nodes.

(5) NTRAC: If the traction boundary conditions over elements p, q, r are the same, then these elements belong to the same "traction group."

(6) NDISP: If the displacement boundary conditions over nodes r, s, t are the same, then these nodes belong to the same "displacement group."

(7) Set KSYM= 0, if no symmetry conditions invoked. See Section 5.4.1 for definition of KSYM, if symmetry about one (or more) planes is required.

(8) NAUTO may be positive, zero, or negative depending on the desired integration strategy. See Section 5.4.1 for definition of NAUTO.

(9) TOLGP: For routine purposes, a tolerance of 10^{-3} may be sufficient, but for highly accurate work, tolerances of less than 10^{-4} may be required. This parameter may take positive or negative values. If positive, the analytical error bound is employed; if negative, the empirical error bound is employed.

(10) I: Nodes must be numbered sequentially (i.e., 1 – NTP), beginning with the "basic" boundary nodes, then the interior nodes (if any), followed by any additional "traction-discontinuous" nodes. Of course, the coordinates of the latter must be identical to the coordinates of the corresponding "basic" nodes.

(11) NFIXU: Refers to "basic" nodes only. The same fixity (prescribed displacement) is applied automatically to the corresponding "traction-discontinuous" nodes (if there are any). Constrained nodes also include those where the prescribed displacements are nonzero.

(12) NUGRP: If node J is constrained by a displacement of type (group) N, then set NUGRP(J) = N.

(13) KODP: Binary numbers (e.g., 101, 011 in three dimensions) that define the displacement constraints (e.g., 101 implies prescription of the x and z displacements). Leading zeros may be omitted. KODP permits the use of one, some, or all specified displacements from a selected displacement group.

(14) MULTP: In an elastic analysis, set this parameter to zero.

(15) The additional nodes K must be numbered sequentially starting from the last node

number for the "basic" boundary nodes and interior nodes (if any). This is illustrated in Figure H.1, in which the total number of nodes (NTP) is ten. Of these, eight nodes (1–8) are "basic" boundary nodes, while node 9 is an interior node and node 10 is an "additional" node resulting from the traction-discontinuity at node 6. Thus, nodes 6 and 10 share the same coordinates but are part of different boundary elements and interior cells.

(16) KBU: Node number in the original list of nodes. (Programmers will note that KBU is also used for another purpose within the program code).

(17) LNDB and LNDC: See Appendix B for node-numbering convention. Elements and cells should be numbered serially (1 – NBE, 1 – NCELL). The number of nodes per boundary element (and per side of each cell) must be equal to NODE. For boundary elements (LNDB), nodes should be numbered in the counterclockwise direction when viewed from outside the domain. If "additional" nodes have been defined to deal with traction discontinuities (using Card Sets 7 and 8), then these nodes should be assigned to the relevant elements (and cells) here.

(18) IFLAG is the traction equivalent to KODP (described in Note 13 above). One difference is that tractions can also be specified in local coordinates (that is, in the tangential and normal directions). This is signaled by changing the sign of IFLAG (i.e., negative).

(19) Array F: Specify the tractions (t_x, t_y, and t_z) for local node 1, then local node 2, etc. Those not prescribed should be set to zero. Tractions may also be prescribed in local coordinates (ξ_1, ξ_2, and n) if IFLAG is negative. The tangential directions (ξ_1, ξ_2) are defined by the node-numbering scheme of Appendix B.

(20) Array RU: Specify the displacements (u_x, u_y, and u_z) for a node. Those not prescribed (or fixed) should be set to zero.

(21) DIAGV takes the values 0.0, 0.5, and 1.0 for closed, semi-infinite, and infinite regions, respectively.

(22) NCRIT: 1 = Tresca; 2 = Von Mises; 3 = Mohr–Coulomb; 4 = Drucker–Prager.

(23) UNIAX: Uniaxial initial yield stress σ_y for Tresca and Von Mises criteria but cohesion c for Mohr–Coulomb and Drucker–Prager criteria.

(24) FRICT: Friction angle ϕ (degrees) for Mohr–Coulomb and Drucker–Prager criteria. Not used by other criteria (set to zero).

(25) HARD(I/K): Hardening moduli are discussed in Section 7.5.3 and the following sections. Perfect plasticity can be recovered by setting both HARDI and HARDK equal to zero.

(26) EM: (0 = kinematic → mixed → isotropic = 1)

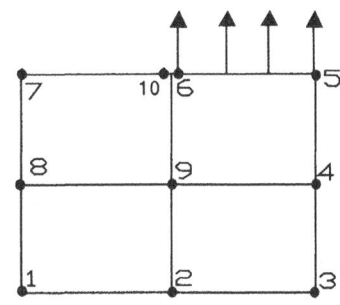

Figure H.1 Numbering of additional nodes.

(27) NINCS: Typically, ten to forty increments are sufficient. Convergence may be checked by experimenting with a range of values.

(28) FINCR: For equal-sized increments, set this parameter equal to unity. Otherwise, set this parameter to less than unity (typically 0.9). A value closer to unity is advisable if a large number of increments is employed.

(29) MITER: Typically, set this to 5. If, in any increment, the actual number of iterations equals MITER, the user should investigate whether the convergence tolerance is too tight (small) or whether some physical instability (i.e., collapse) has occurred. Otherwise, increase MITER.

(30) TOLER: For routine purposes, a convergence tolerance of 10^{-2} may be sufficient, but for highly accurate work a tolerance of less than 10^{-4} may be required. An unreasonably small tolerance value will result in excessive computational time (large number of iterations) without materially affecting the output data.

(31) NOUT1: A four-digit integer number that switches output on/off, during the incremental process. For example, 3303 prints (a) displacements, (b) stresses, and (d) equivalent plastic strains and stresses – every third increment. See Section 10.4 for further details.

(32) NOUT2: A four-digit integer number that switches output on/off, for the last increment (final results) only. For example, 1101 prints (a) displacements, (b) stress, and (d) equivalent plastic strains and stresses. In this example, the zero in the third position switches off printing of (c) back stresses. See Section 10.4 for further details.

References

Aliabadi, M. H. (1997). Boundary element formulations in fracture mechanics, *Appl. Mech. Rev.*, 50, 83–96.

Banerjee, P. K. (1994). *The Boundary Element Methods in Engineering*, McGraw-Hill, London.

Banerjee, P. K. & Butterfield, R. (1981). *Boundary Element Methods in Engineering Science*, McGraw-Hill, Maidenhead, UK.

Banerjee, P. K. & Cathie, D. N. (1980). A direct formulation and numerical implementation of the boundary element method for two-dimensional problems of elastoplasticity, *Int. J. Mech. Sci.*, 22, 233–245.

Banerjee, P. K., Cathie, D. N. & Davies, T. G. (1979). Two and three-dimensional problems of elasto-plasticity. In: *Developments in Boundary Element -1*, P. K. Banerjee & R. Butterfield (Eds.), Elsevier Applied Science Publishers, London.

Banerjee, P. K. & Davies, T. G. (1979). Analysis of some case histories of laterally loaded pile groups. In: *Proc. of Int. Conf. on Numerical Methods in Offshore Piling*, Institution of Civil Engineers, London.

Banerjee, P. K. & Davies, T. G. (1984). Advanced implementation of the boundary element method for three-dimensional problems of elasto-plasticity. In: *Developments in Boundary Element – 3*, P. K. Banerjee & S. Mukherjee (Eds.), Elsevier, London.

Banerjee, P. K. & Driscoll, R. M. C. (1976). Three-dimensional analysis of raked pile-groups, *Proc. Institution of Civil Engineers*, Part 2, 61, 653–671.

Banerjee, P. K., Henry, D. P. & Raveendra, S. T. (1989). Advanced inelastic analysis of solids by the boundary element method, *Int. J. Mech. Sci.*, 31, 309–322.

Becker, A. A. (1992). *The Boundary Element Method in Engineering*, McGraw-Hill, London.

Beer, G. & Watson, J. O. (1989). Infinite boundary elements, *Int. J. Numerical Methods Eng.*, 28, 1233–1247.

Benjumea, R. & Sikarskie, D. L. (1972). On the solution of plane, orthotropic elasticity problems by an integral method, *J. Appl. Mech.*, 39, 801–808.

Blandford, G. E., Ingraffea, A. R. & Liggett, J. A. (1981). Two-dimensional stress intensity factor computations using the boundary element method, *Int. J. Numerical Methods Eng.*, 17, 387–404.

Bonnet, M. & Mukherjee, S. (1996). Implicit BEM formulations for usual and sensitivity problems in elasto-plasticity using the consistent tangent operator concept, *Int. J. Solids Structures*, 33, 4461–4480.

Brebbia, C. A. (1978). *The Boundary Element Method for Engineers*. Pentech Press, London.

Brebbia, C. A. & Dominguez, J. (1992). *Boundary Elements: An Introductory Course*, McGraw-Hill, London.

Brebbia, C. A., Telles, J. C. F. & Wrobel, L. C. (1984). *Boundary Element Techniques*, Springer-Verlag, Berlin.

Bu, S. & Davies, T. G. (1995). Effective evaluation of non-singular integrals in 3D BEM, *Adv. Eng. Software*, 23, 121–128.

Bui, H. D. (1978). Some remarks about the formulation of three-dimensional thermo-elastoplastic problems by integral equations, *Int. J. Solids Structures*, 14, 935–939.

Burke, W. L. (1985). *Applied Differential Geometry*, Cambridge University Press, Cambridge.

Casey, J. & Naghdi, P. M. (1981). On the characterization of strain hardening in plasticity, *J. Appl. Mech.*, 48, 285–296.

Chaudonneret, M. (1978). On the discontinuity of the stress vector in the boundary integral equation method for elastic analysis, In: *Recent Advances in Boundary Element Methods*, C. A. Brebbia (Ed.), Pentech Press, London, 185–94.

Chen, W. F. (1994). *Constitutive Equations for Engineering Materials. Volume 2: Plasticity and Modeling*, Elsevier, London.

Chen, H., Wang, Y. C. & Lu, P. (1996). Stress rate integral equations of elastoplasticity, *ACTA Mechanica Sinica* (English Series), 12, 55–64.

Chopra, M. B. & Dargush, G. F. (1994). Development of BEM for thermoplasticity, *Int. J. Solids Structures*, 31, 1635–1656.

Crisfield, M. A. (1997). *Non-Linear Finite Element Analysis of Solids and Structures*, Wiley, Chichester, UK.

Crotty, J. M. (1982). A block equation solver for large unsymmetric matrices arising in the boundary element method, *Int. J. Numerical Methods Eng.*, 18, 997–1017.

Crouch, S. L. & Starfield, A. M. (1983). *Boundary Element Methods in Solid Mechanics*, Allen & Unwin, London.

Cruse, T. A. (1969). Numerical solutions in three-dimensional elastostatics, *Int. J. Solids Structures*, 5, 1259–1274.

Cruse, T. A. (1974). An improved boundary-integral equation method for three-dimensional elastic stress analysis, *Computers and Structures*, 4, 741–754.

Dallner, R. & Kuhn, G. (1993). Efficient evaluation of volume integrals in boundary element method, *Comput. Methods Appl. Mech. Eng.*, 109, 95–109.

Davies, T. G. & Bu, S. (1996). Infinite boundary elements for the analysis of halfspace problems, *Computers and Geotechnics*, 19, 137–151.

Deb, A., Henry, D. & Wilson, R. B. (1991). An alternate BEM for 2D and 3D anisotropic thermoelasticity, *Int. J. Solids Structures*, 27, 1721–1738.

Dominguez, J. (1993). *Boundary Elements in Dynamics*, Computational Mechanics, Southampton, UK.

Drucker, D. C. (1959). A definition of stable inelastic material, *J. Appl. Mech.* 26, Trans. ASME, Series E, 101.

Drucker, D. C. & Prager, W. (1952). Soil mechanics and plastic analysis or limit design, *Q. Appl. Math.*, 10, 157–165.

Fox, E. N. (1948). The mean elastic settlement of a uniformly loaded area at a depth below the ground surface. In: *Proc. 2nd Int. Conf. on Soil Mechanics and Foundation Engineering*, 1, 129.

Fredholm, I. (1905). Solution d'un probleme fondamental de la theorie de l'elasticitie. *Arkiv Mat. Astron. Fys.*, 2, 1–8.

Fung, Y. C. (1965). *Foundations of Solid Mechanics*, Prentice-Hall, Englewood Cliffs, New Jersey.

Gao, X. W. (1999). "3D Non-Linear and Multi-Region Boundary Element Stress Analysis," PhD thesis, University of Glasgow, UK.

Gao, X. W. & Davies, T. G. (1998). 3-D infinite boundary elements for half-space problems, *Eng. Analysis with Boundary Elements*, 21, 207–213.

Gao, X. W. & Davies, T. G. (2000a). 3D multi-region BEM with corners and edges, *Int. J. Solids Structures*, 37, 1549–1560.

Gao, X. W. & Davies, T. G. (2000b). An effective boundary element algorithm for 2D and 3D elastoplastic problems, *Int. J. Solids Structures*, 37, 4987–5008.

Gao, X. W. & Davies, T. G. (2000c). Adaptive algorithm in elasto-plastic boundary element analysis, *J. Chinese Inst. Eng.*, 23, 349–356.

Gao, X. W. & Lu, J. T. (1992). A combination method of FEM and BEM for elastoplastic problems. In: *Proc. of Fourth Int. Conf. on EPMESC*, Dalian University of Technology Press, China, 630–35.

Gao, X. W. & Zhong, Z. Q. (1992). Elastoplastic damage theory in isotropic medium, *Chinese J. Theoretical Appl. Mech.*, No. 4, 81–91.

Guiggiani, M. & Gigante, A. (1990). A general algorithm for multidimensional Cauchy principal value integrals in the boundary element method, *J. Appl. Mech.*, 57, 906–915.

Guiggiani, M., Krishnasamy, G., Rudolphi, T. J. & Rizzo, F. J. (1992). General algorithm for the numerical solution of hyper-singular boundary integral equations, *(ASME) J. Appl. Mech.*, 59, 604–614.

Hill, R. (1950). *The Mathematical Theory of Plasticity*, Clarendon Press, Oxford.

Huber, O., Lang, A. & Kuhn, G. (1993). Evaluation of the stress tensor in 3D elastostatics by direct solving of hypersingular integrals, *Comput. Mech.*, 12, 39–50.

Huber, O., Dallner, R., Partheymuller, P. & Kuhn, G. (1996). Evaluation of the stress tensor in 3-D elastoplasticity by direct solving of hypersingular integrals, *Int. J. Numerical Methods Eng.*, 39, 2555–2573.

Il'iushin, A. A. (1961). On the postulate of plasticity, *Prikl. Mat. Meh.*, 25, 503–507.

Jaswon, M. A. & Symm, G. T. (1977). *Integral Equation Methods in Potential Theory and Elastostatics*, Academic Press, London.

Kane, J. H. (1994). *Boundary Element Analysis in Engineering Continuum Mechanics*, Prentice-Hall, Englewood Cliffs, New Jersey.

Kane, J. H., Kumar, B. L. K. & Saigal, S. (1990). An arbitrary condensing, noncondensing solution strategy for large scale, multi-zone boundary element analysis, *Comput. Methods Appl. Mech. Eng.*, 79, 219–244.

Lachat, J. C. (1975). "A Further Development of the Boundary Integral Technique for Elastostatics," PhD thesis, University of Southampton, UK.

Lachat, J. C. & Watson, J. O. (1975). A second generation boundary integral program for three dimensional elastic analysis. In: *Boundary Integral Equation Method: Computational Applications in Applied Mechanics*, T. A. Cruse & F. J. Rizzo (Eds.), Vol. 11, ASME, New York, 85–100.

Lachat, J. C. & Watson, J. O. (1976). Effective numerical treatment of boundary integral equation, *Int. J. Numerical Methods Eng.*, 10, 991–1005.

Lade, P. V. (1977). Elasto-plastic stress–strain theory for cohesionless soil with curved yield surfaces, *Int. J. Solids Structures*, 13, 1019–1035.

Lekhnitskii, S. G. (1963). *Theory of Elasticity of an Anisotropic Elastic Body*, J. J. Brandstatter (Ed.), translation by P. Fern, Holden-Day, San Francisco.

Leung, C. Y. & Walker, S. P. (1997). Iterative solution of large three-dimensional BEM elastostatic analyses using the GMRES technique, *Int. J. Numerical Methods Eng.*, 40, 2227–2236.

Linkens, D. (1993). *Selected Benchmarks for Material Non-Linearity*, National Association for Finite Element Methods and Standards, Report R0026, East Kilbride, UK.

Love, A. E. H. (1892). *A Treatise on the Mathematical Theory of Elasticity*, Cambridge University Press, Cambridge. Reprinted New York: Dover Publications, 1963, quoting, Somigliana, C. (1885). Sopra l'equilibrio di un corpo elastico isotropo, *Il Nuovo Cimento*, 3, 17–20.

Lubliner, J. (1990). *Plasticity Theory*, Macmillan, New York.

Massonet, C. E. (1965). Numerical use of integral procedures. In: *Stress Analysis – Recent Developments in Numerical and Experimental Procedures*, O. C. Zienkiewicz & G. S. Hollister (Eds.), Wiley, New York, 198–235.

Mendelson, A. & Albers, L. V. (1975). An application of the boundary integral equation method to elastoplastic problems. In: *Proc. ASME Conf. on Boundary Integral Equation Methods*, T. A. Cruse & F. J. Rizzo (Eds.), ASME, New York.

Mi, Y. & Aliabadi M. H. (1996). A Taylor expansion algorithm for integration of 3D near-singular integrals, *Commun. Numerical Methods Eng.*, 12, 51–62.

Mikhlin, S. G. (1965). *Multi-Dimensional Singular Integrals and Integral Equations*, Pergamon Press, Oxford.

Mukherjee, S. (1977). Corrected boundary integral equations in planar thermo-elasto-plasticity, *Int. J. Solids Structures*, 13, 331–335.

Mustoe, G. G. W. (1984). Advanced integration schemes over boundary elements and volume cells for two- and three-dimensional non-linear analysis. In: *Developments in Boundary Element Methods – 3*, P. K. Banerjee & S. Mukherjee (Eds.), Elsevier, London, 213–270.

Naghdi, P. M. & Trapp, J. A. (1975a). The significance of formulating plasticity theory with reference to loading surfaces in strain space, *Int. J. Eng. Sci.*, 13, 785–797.

Naghdi, P. M. & Trapp, J. A. (1975b). Restrictions on constitutive equations of finite de-formed elastic–plastic materials, *Q. J. Mech. Appl. Math.*, XXVIII, 25–46.

Okada, H. & Atluri, S. N. (1994). Recent developments in the field-boundary element method for finite/small strain elastoplasticity, *Int. J. Solids Structures*, 31, 1737–1775.

Opriessnig, G. & Beer, G. (1999). Visualisation of simulation results for geotechnical prob-lems. In: *Proc. 7th Int. Symposium on Numerical Models in Geomechanics – NUMOG VII*, G. N. Pande, S. Pietruszczak & H. F. Schweiger (Eds.), Balkema, Graz, 353–358.

Ortiz, M. & Simo, J. C. (1986). An analysis of a new class of integration algorithms for elastoplastic constitutive relations, *Int. J. Numerical Methods Eng.*, 23, 353–366.

Ottosen, N. S. (1977). A failure criterion for concrete, *J. Eng. Mech. Division, ASCE*, 103, 527–535.

Owen, D. R. J. & Hinton, E. (1980). *Finite Elements in Plasticity: Theory and Practice*, Pineridge Press Limited, Swansea, UK.

Pan, Y. C. & Chou, T. W. (1976). Point force solution for a transversely isotropic solid, *J. Appl. Mech. (ASME)*, 98, 608–612.

Poon, H., Mukherjee, S. & Bonnet, M. (1998). Numerical implementation of a CTO-based implicit approach for the BEM solution of usual and sensitivity problems in elasto-plasticity, *Eng. Analysis with Boundary Elements*, 22, 257–269.

Poulos, H. G. & Davis, E. H. (1974). *Elastic Solutions for Soil and Rock Mechanics*, Wiley, New York.

Press, W. H., Teukolsky, S. A., Vetterling, W. T. & Flannery, B. P. (1992). *Numerical Recipes in FORTRAN: The Art of Scientific Computing*, 2nd ed., Cambridge University Press, New York.

Raveendra, S. T. (1984). "Advanced Development of BEM for Two and Three-Dimensional Inelastic Analysis," PhD thesis, State University of New York at Buffalo.

Raveendra, S. T. & Banerjee, P. K. (1992). Eigenvalue analysis by boundary element method. In: *Advanced Dynamic Analysis by Boundary Element Methods*, P. K. Banerjee & S. Kobayashi (Eds.), Elsevier Applied Science, London, 282–320.

Riccardella, P. (1973). "An Implementation of the Boundary Integral Technique for Planar Problems of Elasticity and Elastoplasticity," PhD thesis, Carnegie Mellon University, Pittsburgh, Pennsylvania.

Rizzo, F. J. (1967). An integral equation approach to boundary value problems of classical elastostatics, *Q. J. Appl. Math.*, 25, 83–95.

Saad, Y. & Schultz, M. H. (1986). GMRES: A generalized minimal residual algorithm for solving non-symmetric linear systems, *SIAM J. Sci. Statistical Computing*, 7, 856–869.

Simo, J. C. & Govindjee, S. (1991). Non-linear B-stability and symmetry preserving return mapping algorithms for plasticity and viscoplasticity, *Int. J. Numerical Methods Eng.*, 31, 151–176.

Simo, J. C. & Taylor, R. L. (1985). Consistent tangent operators for rate-independent elasto-plasticity, *Comput. Methods Appl. Mech. Eng.*, 48, 101–118.

Sladek, V. & Sladek, J. (1998). *Singular Integrals in Boundary Element Methods*, WIT Press, Southampton, UK.

Smith, I. M. (1995). *Programming in FORTRAN90 – A First Course for Engineers and Scientists*, Wiley, London.

Snyder, M. D. & Cruse, T. A. (1975). Boundary integral equation analysis of cracked anisotropic plates, *Int. J. Fracture*, 11, 315–328.

Somigliana, C. (1885). Sopra l'equilibrio di un corpo elastico isotropo, *Il Nuovo Cimento*, 3, 17–20, quoted by, Love, A. E. H. (1982). *A Treatise on the Mathematical Theory of Elasticity*, Cambridge University Press, Cambridge. Reprinted New York: Dover Publications, 1963.

Southwell, R. V. (1946). *Relaxation Methods in Theoretical Physics*, Oxford University Press, Oxford, UK.

Stroud, A. H. & Secrest, D. (1966). *Gaussian Quadrature Formulas*, Prentice-Hall, Englewood Cliffs, New Jersey.

Swedlow, J. L. & Cruse, T. A. (1971). Formulation of boundary integral equations for three-dimensional elasto-plastic flow, *Int. J. Solids Structures*, 7, 1673–1683.

Telles, J. C. F. (1983). *The Boundary Element Method Applied to Inelastic Problems*, Springer-Verlag, Berlin.

Telles, J. C. F. & Brebbia, C. A. (1979). On the application of the boundary element method to plasticity, *Appl. Math. Modelling*, 3, 466–470.

Telles, J. C. F. & Brebbia, C. A. (1980). The boundary element method in plasticity. In: *Proc. Second Int. Conf. on Recent Advances in Boundary Element Methods*, C. A. Brebbia (Ed.), Pentech Press, Plymouth, UK, 295–317.

Telles, J. C. F. & Carrer, J. A. M. (1991). Implicit procedures for the solution of elasto-plastic problems by the boundary element method, *Math. Comput. Modelling*, 15, 303–311.

Thomson, W. (Lord Kelvin) (1848). A note on the integration of the equations of equilibrium of an elastic solid, *Cambridge and Dublin Mathematical J.* February. (Reproduced in *Physical & Mathematical Papers*, 1, 97–99 (W. Thomson, Ed.), Cambridge University Press, 1882).

Timoshenko, S. P. & Goodier J. N. (1970). *Theory of Elasticity*, McGraw-Hill, New York.

Tomlin, G. R. & Butterfield, R. (1974). Elastic analysis of zoned orthotropic media, *J. Eng. Mech. Division, ASCE*, 100, 511–529.

Watson, J. O. (1979). Advanced implementation of the boundary element method for two- and three-dimensional elastostatics. In: *Developments in Boundary Element Methods – 1*, P. K. Banerjee & R. Butterfield (Eds.), Elsevier Applied Science, London, 31–63.

Wearing, J. L. & Dimagiba, R. R. M. (1998). The development of the boundary element method for three dimensional elasto-plastic analysis. In: *Boundary Element Research in Europe*, C. A. Brebbia (Ed.), Computational Mechanics, Southampton, UK, 93–102.

Wilde, A. J. (1998). "A Hypersingular Dual Boundary Element Formulation for Three-Dimensional Fracture Analysis," PhD thesis, Wessex Institute of Technology, University of Wales, UK.

Zhang, Q. & Mukherjee, S. (1991). Design sensitivity coefficients for linear elastic bodies with zones and corners by the derivative boundary element method, *Int. J. Solids Structures*, 27, 983–998.

Ziegler, H. (1959). A modification of Prager's hardening rule, *Q. J. Appl. Math.*, 17, 55–65.

Zienkiewicz, O. C. (1977). *The Finite Element Method*, McGraw-Hill, Maidenhead, UK.

Zienkiewicz, O. C. & Taylor, R. L. (1989). *The Finite Element Method*, McGraw-Hill, Maidenhead, UK.

Index